CLIMATE CHANGE AND ITS CAUSES, EFFECTS AND PREDICTION

CLIMATE CHANGE EFFECTS ON SOILS

ASPECTS AND CONSIDERATIONS

CLIMATE CHANGE AND ITS CAUSES, EFFECTS AND PREDICTION

Additional books in this series can be found on Nova's website under the Series tab.

Additional e-books in this series can be found on Nova's website under the e-book tab.

CLIMATE CHANGE AND ITS CAUSES, EFFECTS AND PREDICTION

CLIMATE CHANGE EFFECTS ON SOILS

ASPECTS AND CONSIDERATIONS

CLAUDIA HOLMES
EDITOR

New York

Copyright © 2015 by Nova Science Publishers, Inc.

All rights reserved. No part of this book may be reproduced, stored in a retrieval system or transmitted in any form or by any means: electronic, electrostatic, magnetic, tape, mechanical photocopying, recording or otherwise without the written permission of the Publisher.

We have partnered with Copyright Clearance Center to make it easy for you to obtain permissions to reuse content from this publication. Simply navigate to this publication's page on Nova's website and locate the "Get Permission" button below the title description. This button is linked directly to the title's permission page on copyright.com. Alternatively, you can visit copyright.com and search by title, ISBN, or ISSN.

For further questions about using the service on copyright.com, please contact:
Copyright Clearance Center
Phone: +1-(978) 750-8400   Fax: +1-(978) 750-4470   E-mail: info@copyright.com.

NOTICE TO THE READER

The Publisher has taken reasonable care in the preparation of this book, but makes no expressed or implied warranty of any kind and assumes no responsibility for any errors or omissions. No liability is assumed for incidental or consequential damages in connection with or arising out of information contained in this book. The Publisher shall not be liable for any special, consequential, or exemplary damages resulting, in whole or in part, from the readers' use of, or reliance upon, this material. Any parts of this book based on government reports are so indicated and copyright is claimed for those parts to the extent applicable to compilations of such works.

Independent verification should be sought for any data, advice or recommendations contained in this book. In addition, no responsibility is assumed by the publisher for any injury and/or damage to persons or property arising from any methods, products, instructions, ideas or otherwise contained in this publication.

This publication is designed to provide accurate and authoritative information with regard to the subject matter covered herein. It is sold with the clear understanding that the Publisher is not engaged in rendering legal or any other professional services. If legal or any other expert assistance is required, the services of a competent person should be sought. FROM A DECLARATION OF PARTICIPANTS JOINTLY ADOPTED BY A COMMITTEE OF THE AMERICAN BAR ASSOCIATION AND A COMMITTEE OF PUBLISHERS.

Additional color graphics may be available in the e-book version of this book.

Library of Congress Cataloging-in-Publication Data

ISBN: 978-1-63482-773-7

Published by Nova Science Publishers, Inc. † New York

CONTENTS

Preface		**vii**
Chapter 1	Overview of Different Aspects of Climate Change Effects on Soils *N. Oafoku*	**1**
Chapter 2	Climate Change Science: Key Points *Jane A. Leggett*	**61**
Index		**81**

PREFACE

This book reports recent discoveries and identifies key research needs required to understand the effects of climate change on soils.

Chapter 1 – Climate change has and will significantly affect soil properties. Biotic processes that consume atmospheric CO_2 and create organic carbon (C) that is either reprocessed to CO_2 or stored in soils, are the subject of active current investigations with great concern over the influence of climate change.

In addition, abiotic C cycling and its influence on the inorganic C pool in soils is a fundamental global process in which acidic atmospheric CO_2 participates in the weathering of carbonate and silicate minerals, ultimately delivering bicarbonate and Ca^{2+} or other cations that precipitate in the form of carbonates in soils or are transported to the rivers, lakes, and oceans.

Soil responses to climate change will be complex, and there are many uncertainties and unresolved issues. The objective of this review is to initiate and further stimulate a discussion about some important and challenging aspects of climate-change effects on soils, such as accelerated weathering of soil minerals and resulting C and elemental fluxes in and out of soils, soil/geo-engineering methods used to increase C sequestration in soils, soil organic matter (SOM) protection, transformation and mineralization, and SOM temperature sensitivity.

This review reports recent discoveries and identifies key research needs required to understand the effects of climate change on soils.

Chapter 2 – Though climate change science often is portrayed as controversial, broad scientific agreement exists on many points:

- The Earth's climate is warming and changing.
- Human-related emissions of greenhouse gases (GHG) and other pollutants have contributed to warming observed since the 1970s and, if continued, would tend to drive further warming, sea level rise, and other climate shifts.
- Volcanoes, the Earth's relationship to the Sun, solar cycles, and land cover change may be more influential on climate shifts than rising GHG concentrations on other time and geographic scales. Human-induced changes are super-imposed on and interact with natural climate variability.
- The largest uncertainties in climate projections surround feedbacks in the Earth system that augment or dampen the initial influence, or affect the pattern of changes. Feedback mechanisms are apparent in clouds, vegetation, oceans, and potential emissions from soils.
- There is a wide range of projections of future, human-induced climate change, all pointing toward warming and associated sea level rise, with wider uncertainties regarding the nature of precipitation, storms, and other important aspects of climate.
- Human societies and ecosystems are sensitive to climate. Some past climate changes benefited civilizations; others contributed to the demise of some societies. Small future changes may bring benefits for some and adverse effects to others. Large climate changes would be increasingly adverse for a widening scope of populations and ecosystems.

As is common and constructive in science, scientists debate finer points. For example, a large majority but not all scientists find compelling evidence that rising GHG have contributed the most influence on global warming since the 1970s, with solar radiation a smaller influence on that time scale. Most climate modelers project important impacts of unabated GHG emissions, with low likelihoods of catastrophic impacts over this century. Human influences on climate change would continue for centuries after atmospheric concentrations of GHG are stabilized, as the accumulated gases continue to exert effects and as the Earth's systems seek to equilibrate.

The U.S. government and others have invested billions of dollars in research to improve understanding of the Earth's climate system, resulting in major improvements in understanding while major uncertainties remain. However, it is fundamental to the scientific method that science does not provide absolute proofs; all scientific theories are to some degree provisional and may be rejected or modified based on new evidence. Private and public decisions to act or not to act, to reduce the human contribution to climate change or to prepare for future changes, will be made in the context of accumulating evidence (or lack of evidence), accumulating GHG concentrations, ongoing debate about risks, and other considerations (e.g., economics and distributional effects).

In: Climate Change Effects on Soils
Editor: Claudia Holmes

ISBN: 978-1-63482-773-7
© 2015 Nova Science Publishers, Inc.

Chapter 1

OVERVIEW OF DIFFERENT ASPECTS OF CLIMATE CHANGE EFFECTS ON SOILS[*]

N. Oafoku

1.0. INTRODUCTION

Climate change that is defined by high atmospheric carbon dioxide (CO_2) concentrations (≥400 ppm); increasing air temperatures (2-4°C or greater); significant and/or abrupt changes in daily, seasonal, and inter-annual temperature; changes in the wet/dry cycles; intensive rainfall and/or heavy storms; extended periods of drought; extreme frost; and heat waves and increased fire frequency, is expected to significantly impact terrestrial systems, soil properties, surface water and stream-flow (Patterson et al. 2013); groundwater quality, water supplies and terrestrial hydrologic cycle (Pangle et al. 2014); and, as a consequence, food security and environmental quality. Increased global CO_2 emissions, estimated at 8.4 Pg carbon (C) yr^{-1} in 2010, have accelerated from 1% yr^{-1} during 1990–1999 to 2.5% yr^{-1} during 2000–2009 (Friedlingstein et al. 2010), being the main driver of the global warming. Climate-change impacts, which already are being felt in agriculture, ecosystems, and forests, are expected to be diverse and complex.

---

[*] This is an edited, reformatted and augmented version of a report, PNNL-23483, prepared by Pacific Northwest National Laboratory under contract with the U.S. Department of Energy, August 2014.

The C cycle, the process by which C circulates between the atmosphere, oceans, and lithosphere, includes fossil fuel deposits and the biosphere of the earth (Falkowski et al. 2000). Terrestrial landscapes and ecosystems play a large role in the global C cycling. More than 75,000,000 Gt of C is present within the global C lithosphere pool; by far, the largest quantity of C, more than 60,000,000 Gt, is in the form of sedimentary carbonates, relative to the other C pools (i.e., 720 Gt in the atmosphere, 38,400 Gt in the oceans) (Falkowski et al. 2000). While much of the lithospheric inorganic C (IC) is currently considered to be unavailable to C cycling, abiotic processes associated with climate-induced soil acidification and accelerated mineral weathering could redistribute large quantities of C among the three major Earth's C pools (land, atmosphere and oceans) by significantly altering the C source/sink behavior of soils.

World soils, a large reservoir of reactive C, moderate the global C cycle and atmospheric chemistry (Lal 2013). The soil C pool includes the organic carbon (OC) (living organisms and organic compounds) and the IC (Prentice 2001; Rice 2002). About 2500 Gt C is stored in soils out of which 1550 Gt are OC and 950 Gt are IC (Lal 2004). Clearly, a substantial amount of C in soils is present in the form of carbonates, such as calcite (Lal 2008) and this IC pool may substantially contribute to the global C cycle under conditions of a changing climate.

The C cycle is dynamic and responsive to climate change (Lal 2013). The effects of non-uniform climate warming (Xia et al. 2014), and a variety of human activities have already altered the terrestrial chemical cycles and land-ocean flux of major elements, although the extent remains difficult to quantify (Hartmann et al. 2013). In addition, our current knowledge is insufficient to describe the interactions among the components of the earth system and the relationship between the C cycle and other biogeochemical and climatological processes (Falkowski et al. 2000; Tagliabue et al. 2014). Among uncertainties are fluxes in and out of soils and interfaces (emission and sequestration); the efficiency of natural soil sinks and factors controlling source behavior; mineral silicate, and carbonate weathering and carbonation; SOM protection, (soil organic matter) transformation and mineralization; the fate of eroded dissolved inorganic C (DIC) and dissolved organic C (DOC); and the stability of IC and OC sequestered in soils.

Climate change may also induce a suite of not well understood effects in soils all of which have the potential to affect and/or even control the C balance and elemental cycling in soils. Soil-related consequences include significant/dramatic changes in soils properties, surface water and groundwater quality, food (national) security, water supplies, human health, energy, agriculture, forests, and ecosystems. However, the impact and consequences of climate-change variables on relevant reactions and processes occurring within the surface layer of the terrestrial systems (i.e., soil system) and those occurring at the soil-atmosphere and soil-subsoil interfaces currently are not well understood. In addition, extreme events caused by climate change (Angélil et al. 2014) may have long-term effects on soils with poorly understood consequences.

Soil response to climate change is expected to be multifaceted and rather complicated because of 1) the presence of an intricate network of sequential, simultaneous and/or coupled (often, time-dependent) chemical, biological and hydrological reactions and processes; 2) chemical elements, nutrients, and contaminants involved in these reactions and processes are distributed in the soil solid, liquid, and gas phases; 3) the scale-dependent effects related to mineralogical, chemical, and physical heterogeneities; and 4) climate extremes (e.g., heat waves and dry spells) induce interconnected short- and long-lasting effects in soils that currently are not well understood. To address these and other related issues mentioned above, studies should be conducted at different spatial scales [molecular, nano, soil particle (μm), soil aggregate (mm), soil horizon (cm), soil type, soil order, regional, and global] and temporal scales (minutes to day, days to year, years to decade, decades to century, centuries to millennium, and millennia to epoch/era).

The objective of this review is to initiate and further stimulate discussion about climate-change effects on soils; present the state-of-the-science in important topics related to C and elemental cycling and SOM (soil organic matter) role in fate; and provide ideas for future research directions and needs based on research results and recommendations derived from recently published articles mainly in high visibility journals. In this report, the discussion will be focused mainly on important and challenging aspects of climate-change effects on soils, such as climate-induced accelerated weathering of soil minerals; SOM protection, transformation and mineralization; SOM temperature sensitivity; and C and elemental cycling in soils. This review reports recent discoveries and identifies key research needs required to understand the effects of climate change on soils.

2.0. CLIMATE-CHANGE INDUCED ACCELERATED SOIL-MINERAL WEATHERING AND C CYCLING

2.1. Accelerated Mineral Weathering

Interest in soil-mineral weathering has increased over recent years because of the possible effects of climate change on soil properties and environmental quality and food security; the role soils play in controlling global C cycle; and the positive or negative feedback to a warming climate. The weathering of alkaline rocks, such as alkaline or alkaline earth silicates, is thought to have played an important role in the historical reduction of the atmospheric CO_2 (Kojima et al. 1997), and will have an important role in the evolution of the global C cycle over the next century (Beaulieu et al. 2012), when climate change is expected to be significant.

Accelerated weathering of the rocks and minerals in soils will be promoted by higher atmospheric CO_2 concentrations (≥ 400 ppm) and temperature (which increase the extent and rates of weathering), intensive rainfall (which facilitates the removal of reaction products either by surface runoff or percolating water), and heat waves and extended periods of drought (which promote physical alteration of rocks and minerals). The results from a 44-year field study show that weathering rates are already increasing because of global warming (Gislason et al. 2009). However, the spatial patterns, temporal trends, and controlling factors of the processes and reactions and their effects on different scales, especially regional, continental, and global scales, are not fully understood at this time (Moosdorf et al. 2011).

In addition, although the effects of weathering, intensive leaching and soil acidification have been studied in the past [e.g., studies conducted with highly weathered tropical and subtropical soils (Qafoku et al. 2004; Qafoku et al. 2000; Bellini et al. 1996; Fiantis et al. 2009)], many positive and negative effects of climate-change induced accelerated weathering are not well studied and understood. On the positive side, weathering has the potential to increase the IC pool in soils via carbonate mineral formation, thus contributing to decreasing atmospheric CO_2 concentration. In addition, dissolution promoted release of elements that serve as nutrients for microbes and plants may stimulate microbial activity and increase plant productivity and biotic C sequestration, which will further increase C removal from the atmosphere. On the negative side, the acceleration of weathering may perturb the balance of processes not only in the biotic C cycle but also in the abiotic C cycles within soils, thereby controlling and/or affecting the distribution of C into less stable soil pools; increasing contaminant mobilization that may significantly alter soil microbial activity, plant productivity, life in soils, and C and elemental cycling; and possibly changing in a significant way the elemental balances in rivers, lakes, and oceans.

2.1.1. Soil Acidification and Weathering

The dissolution of atmospheric CO_2 gas in soil water and the subsequent formation of carbonic acid followed by its dissociation cause a decrease in soil pore water pH as a result of aqueous phase proton enrichment through the following chemical reaction:

$$CO_2(g) + H_2O \leftrightarrow H_2CO_3 \leftrightarrow HCO_3^-(aq) + H^+(aq)$$

Experimental and modeling studies conducted with soil and subsoil materials have shown a decrease in aqueous pH of 1 to 3 units in soil pore water as a result of excess exposure to CO_2 gas (Altevogt and Jaffe 2005; Little and Jackson 2010; Lu et al. 2010; Vong et al. 2011; Wang and Jaffe 2004; Wei et al. 2011; Wilkin and DiGiulio 2010; Zheng et al. 2009; Kharaka et al. 2010). A decrease in pH on the lower end of this range is typical of well-buffered systems in which CO_2-induced dissolution of reactive carbonates, silicates, and clay minerals provides enough buffering capacity (via HCO_3^- alkalinity) to resist changes in pH. Poorly buffered systems (e.g., sandy soils) have low abundance or are devoid of alkalinity-producing minerals and, therefore, lack the ability to resist changes in pH. In such systems, the decrease in pH is generally more pronounced and may have long-term consequences, and the risk for pH-induced perturbation of these systems is more significant compared to well-buffered systems (McGrail et al. 2006; Wilkin and DiGiulio 2010; Wang and Jaffe 2004).

The following reactions describe the dissolution of soil minerals such as calcite, feldspar and a typical 1:1 phyllosilicate in the presence of an excess amount of CO_2 gas [(Harvey et al., 2013) and the references cited therein]:

$$CaCO_3 + CO_2(g) + H_2O \rightarrow Ca^{2+} + 2HCO_3^-$$

$$2NaAlSi_3O_8 + 11H_2O + 2CO_2(g) \rightarrow Al_2Si_2O_5(OH)_4 + 2Na^+ + 2HCO_3^- + 4H_4SiO_4$$

$$Al_2Si_2O_5(OH)_4 + 5H_2O + 6CO_2(g) \rightarrow 2Al^{3+} + 6HCO_3^- + 2H_4SiO_4$$

The dissolution rate of a rock or mineral is likely to be a function of its composition, crystallinity, composition and temperature of the aqueous phase (Gislason et al. 2010), the external surface roughness and internal porosity (reactive surface area), occurrence as single particles or in aggregates, and location within aggregate structure. Among the most important soil-related controls on the extent and rate of dissolution are soil mineralogy (i.e., minerals that are more or less resistant to weathering), soil type (i.e., highly weathered or less weathered soils), calcite content (i.e., soils with appreciable amounts of calcite with higher buffering capacity to resist changes in pH or soil with less or no calcite), soil texture (i.e., sandy or clay soils) and internal and external reactive surface area. It follows, therefore, that a broad array of soil responses to changes in climate related variables is quite possible.

2.1.2. Accelerated Weathering (The Case of Mine-Tailings and Mountain Denudation)

Publications on accelerated weathering of soils induced by climate change are not present in the current literature. However, papers covering the related topic of the impact of climate change on mine-tailings weathering and, more recently, mountain weathering and erosion have been published recently. Weathering of mine-tailings is expected to proceed more rapidly under global warming because of the dual effects of climate-change variables on weathering and increased acidity generated during weathering, which further accelerates mineral weathering (Wilson et al. 2006; Wilson et al. 2009; Wilson et al. 2010; Mills et al. 2010; Wilson et al. 2011; Bea et al. 2012; Tripathi et al. 2014). A positive effect of mine-tailings weathering is atmospheric CO_2 sequestration via carbonation of mine-tailings (an in-depth discussion about this topic is included in another section of this report). On the other hand, more rapid weathering of mine-tailings may increase contaminant mobilization, and this aspect of mine-tailings weathering seems to be overlooked in the current literature.

An important aspect of climate-change induced weathering in mountainous terrestrial systems is the relationship between weathering and erosion. It is widely recognized that these processes are positively correlated across diverse landscapes, although there are limits to this relationship that remain largely untested (Dixon and von Blanckenburg 2012). Using new global data compilations of soil production and weathering rates from cosmogenic nuclides and silicate weathering fluxes from global rivers, Dixon and von Blanckenburg (2012) showed that the weathering-erosion relationship was capped by what they called "speed limits." Their estimates indicated a soil production speed limit of between 320 to 450 t km^{-2} yr^{-1} and an associated weathering rate speed limit of roughly 150 t km^{-2} yr^{-1}. According to the authors, these limits are valid for a range of lithologies and also extend to mountain belts, where soil cover is not continuous and erosion rates outpace soil production rates. The influence of climate on the formation rates of regolith (i.e., the mantle of physically, chemically, and biologically altered material overlying bedrock), which covers much of earth's continents, was investigated in another recent paper (Dere et al. 2013). Studies and research efforts similar to the latter study should be further expanded to better understand future effects of climate-change variables on soil weathering.

2.2. Weathering-Induced C Fluxes

2.2.1. Atmospheric C Consumption during Mineral Weathering and CO_2 Breathing

Atmospheric CO_2 consumption by chemical weathering is the transformation of CO_2 gas into dissolved bicarbonate ion whose negative charge is balanced by cations such as Ca, Mg, Sr, Fe, and Mn released from chemical weathering of carbonate and silicate minerals (Renforth 2012; Cao et al. 2012; Kelemen and Matter 2008; Tomkinson et al. 2013; Kelemen et al. 2011; Thom et al. 2013; Ryskov et al. 2008). The uptake of atmospheric/soil CO_2 by carbonate rock dissolution plays an important role in the global C cycle, as it is one of the most important C sinks (Cao et al. 2012).

It is well stated in the literature that most C on earth is bound within minerals (Wilson, Raudsepp and Dipple 2006), and atmospheric CO_2 consumption as a result of weathering is an important and integral part of the global C cycle (Moosdorf et al. 2011; Li et al. 2014). Soil minerals, such as wollastonite, which is an alkaline silicate mineral, can provide the divalent cation needed to capture atmospheric CO_2 at ambient environmental conditions via the following reaction (Salek et al. 2013):

$$CaSiO_3 + H_2CO_3 + H_2O = CaCO_3 + H_4SiO_4$$

Olivine dissolution is accompanied by the sequestration of 4 moles of CO_2 for each mole of olivine through the following reaction (Köhler et al. 2010): f

$$Mg_2SiO_4 + 4CO_2 + 4H_2O \rightarrow 2Mg^{2+} + 4HCO_3^- + H_4SiO_4$$

Chemical weathering of continental surfaces consumes 0.3 Gt yr^{-1} of atmospheric C (Gaillardet et al. 1999). This flux is of the same general magnitude as the net uptake of CO_2 by the terrestrial biosphere (0.4 Gt C yr^{-1}) in pre-industrial conditions (IPCC Climate Change 2007). Other studies have shown that about 64% of the average bicarbonate flux caused by chemical weathering in North America originates from atmospheric CO_2 and 36% from dissolution of carbonate minerals (Moosdorf et al. 2011). However, the spatial patterns of dissolution and consumption of atmospheric C and the controlling factors at different scales are not well understood. There is a need for a globally representative set of regionally calibrated models of CO_2 consumption by chemical weathering and its subsequent fate, which would apply very detailed spatial data to resolve the heterogeneity of processes occurring at the earth surface (Moosdorf et al. 2011).

A recent study has shown that global IC interactions and feedbacks in the lithospheric and atmospheric interfaces may be even more dynamic than previously thought. Carbonate anions intercalated within layered double hydroxides, a class of hydrotalcite, may undergo an unusual dynamic exchange with carbonate anions derived from atmospheric CO_2 under ambient conditions (Ishihara et al. 2013). The authors show that the rate of exchange is promoted by low relative humidity levels because of the formation of interlayer nanospace vacancies that acted as initial points for CO_2 uptake from air. Because various hydrotalcite-like minerals exist in soils, it is important to determine the extent and rates of the "soil mineral CO_2 breathing" and subsequent "weathering" of structural carbonate under changing climate conditions, especially in response to wet/drying cycles and increasing CO_2 concentrations in the atmosphere.

2.2.2. Inorganic C and Carbonation

Carbonation is the water-mediated replacement of silicate minerals by carbonate minerals in the Earth's (or even Mars's) crust (Kojima et al. 1997; Olsson et al. 2012; Thom et al. 2013; Tomkinson et al. 2013). Carbonation is important in understanding earth's C cycle and mineral weathering (Kelemen and Matter 2008; Kelemen et al. 2011). Soils contain both carbonate remnants from the rocks (i.e., lithogenic carbonates) and pedogenic carbonates formed during soil formation (Ryskov et al. 2008). Apparently, the abundance of pedogenic carbonates depends primarily on the climate condition and, to a lesser extent, on soil age and content of lithogenic clasts in the parent material (Ryskov et al. 2008). These researchers found two epochs of carbonate formation in the soils of European Russia that coincide with periods of arid climate occurring approximately 3750 and 2300 years ago.

Minerals within the soil environment can capture and store atmospheric C through a two-step process: soil-mineral weathering followed by secondary carbonate mineral precipitation (Washbourne et al. 2012). This process provides a unique and efficient mechanism for CO_2 sequestration in surface terrestrial environments. The dissolved CO_2 (and aqueous bicarbonate) in the soil liquid phase is combined with other reactants released during soil-mineral weathering. The increasing activity of weathering reaction products, such as Ca, Mg or Fe, and HCO_3^-, could lead to supersaturation of the soil solution and precipitation of carbonates [e.g., calcite ($CaCO_3$), magnesite ($MgCO_3$), and siderite ($FeCO_3$)]. This is a naturally occurring C sequestration pathway that could contribute to the global-scale efforts to reduce C concentrations in the atmosphere. Studies have shown that, in the course of soil formation over the last 5000 years, the soil fixed atmospheric CO_2 as pedogenic carbonate during arid periods at a rate of 2.2 kg C m^{-2} y^{-1} in chernozem (Typic Calciustoll), 1.13 kg C m^{-2} y^{-1} in dark-chestnut soil (Typic Haplocalcid), and 0.86 kg C m^{-2} y^{-1} in light-chestnut soil (Typic Natrargid) (Ryskov et al. 2008).

The rate of dissolution of silicate minerals is usually the limiting step in the two-step process of soil-mineral dissolution and CO_2 trapping in the newly formed pedogenic carbonate phases (Gislason et al. 2010). The temperature and composition of the contacting solution also would affect the extent and rate of dissolution and precipitation, and it is expected that the response would be dependent on the parent rock type. For example, the higher divalent metal content of ultramafic and mafic rocks would make them better candidates for carbonation than sandstone. Previous research suggests that such compositional variations can result in about two orders-of-magnitude difference in the rate of carbonation (Gislason et al. 2010).

Several studies have demonstrated mineralogical trapping of CO_2 via mineral carbonation under conditions associated with deep geologic CO_2 storage (Loring et al. 2011; Oldenburg and Unger 2003; Bearat et al. 2006; McGrail et al. 2009; Harvey et al. 2013; Birkholzer et al. 2008; Langmuir 1997), but similar assessments have not been considered for climate-change induced mineralogical trapping of atmospheric CO_2 in surface environments (soils included). A full discussion of thermodymanic and kinetic limitations to pedogenic carbonate precipitation in subsoil environments is provided in the review paper by Harvey et al. (2003). While calcite may be readily formed in most soil environments, there are important kinetic limitations to precipitation of some other carbonate minerals (e.g., magnesite, dolomite, and siderite) even in cases where soil solutions are supersaturated with respect to these phases (Saldi et al. 2009; Arvidson and Mackenzie 1999; Jimenez-Lopez and Romanek 2004). This might be explained by the differences in surface charge density between Fe^{2+} and Ca^{2+} or Mg^{2+} and Ca^{2+} ions and the higher activation energy required to initiate dehydration of dissolved Fe^{2+} and Mg^{2+} ions and precipitation of siderite, magnesite, or dolomite. However, precursors such as nesqhehonite ($MgCO_3 \times 3H_2O$) may form in low-temperature CO_2-rich solutions that are supersaturated with respect to magnesite (Loring et al. 2012; Harrison et al. 2013; Beinlich and Austrheim 2012). Atmospheric CO_2 consumption may therefore occur via formation of these precursors in soils, although additional work is needed to study the stability of these phases under different soil and climate-change conditions.

Some authors have noticed that calcite precipitation can be inhibited by natural SOM (soil organic matter) dissolved in solutions (Lin et al. 2005). SOM may sorb to the calcite surface, and the greatest degree of inhibition has been observed for the highest molecular weight and aromatic C content of the materials (Lin et al. 2005). However, the degree to which the SOM/calcite interactions control calcite precipitation under different conditions and are affected by climate-change variables is unknown.

Microbes may play an important role in carbonate formation (Kenward et al. 2009; Power et al. 2011; Roberts et al. 2004). Examples include the formation of dolomite in a basalt aquifer on time-scales of weeks to months mediated by methanogens (Roberts et al. 2004). The rate of carbonation may be increased by carbonate-forming microorganisms that use their cells as a seeding/nucleation surface (Dong 2010; Douglas and Beveridge 1998). The presence of a suitable seeding surface can increase carbonate precipitation rates by up to threefold (Lin and Singer 2005). Microbe-mediated carbonate formation (Mozley and Burns 1993) and generation of bicarbonate ions used in hydrophyte photosynthesis, both processes resulting in IC (i.e., the HCO_3^- ions) being converted to OC (Cao et al. 2012), are other pathways of atmospheric CO_2 removal and transformation in soils that may be important under changing climate conditions. Clearly, elucidating microbially mediated feedbacks is fundamental to understanding ecosystem responses to climate warming, and would help provide a mechanistic basis for C-climate modeling (Zhou et al. 2012).

One last important topic in this section is the possibility of secondary phase formation, other than carbonates, resulting from accelerated soil weathering. Experiments conducted to study olivine weathering demonstrated the formation of goethite, hematite, silica, and carbonate minerals both in the presence or absence of air; formation of Fe^{3+} oxides in the absence of oxygen suggested hydrolysis, where water was converted to hydrogen and oxygen (Olsson et al. 2012). The formation of other secondary phases, in addition to carbonates, is a very important aspect of soil response to climate change that needs attention and further exploration.

2.2.3. Carbonation of Mine-Tailings and Urban Soils

Many recent studies have addressed the topic of CO_2 sequestration via weathering and carbonation of mine-tailings (Wilson, Raudsepp and Dipple 2006; Wilson et al. 2009; Wilson et al. 2010; Wilson et al. 2011; Bea et al. 2012; Mills et al. 2010; Tripathi, Singh and Nathanail 2014). Characterized samples of neophases present in mine-tailings indicated that modern atmospheric CO_2 was used during mineral formation in a soil-like environment with characteristics similar to soil carbonate (Wilson et al. 2009). Carbonate phases include hydrated Mg carbonate minerals such as nesquehonite [$MgCO_3 \times 3H_2O$], dypingite [$Mg_5(CO_3)_4(OH)_2 \times 5H_2O$], hydromagnesite [$Mg_5(CO_3)_4(OH)_2 \times 4H_2O$], and less common lansfordite [$MgCO_3 \times 5H_2O$]. The potential for decomposition of metastable hydrated Mg carbonate phases to geologically stable magnesite represents a possible pathway of long-term mineral CO_2 sequestration in mine-tailings (Wilson, Raudsepp and Dipple 2006). Dypingite also may precipitate out of high-pH, high-salinity solutions (Wilson et al. 2010).

Other studies have found that atmospheric CO_2 is naturally sequestered in ultramafic mine-tailings as a result of the weathering of serpentine minerals [$Mg_3Si_2O_5(OH)_4$] and brucite [$Mg(OH)_2$], and subsequent mineralization of CO_2 in hydrated magnesium carbonate minerals, such as hydromagnesite [$Mg_5(CO_3)_4(OH)_2 \times 4H_2O$] (Bea et al. 2012). In this study, kinetic dissolution of serpentine, dissolution-precipitation of brucite and primary carbonates- calcite ($CaCO_3$), dolomite [$MgCa(CO_3)_2$], magnesite ($MgCO_3$), as well as the formation of hydromagnesite, halite (NaCl), gypsum ($CaSO_4 \times 2H_2O$), blödite [$Na_2Mg(SO_4)_2 \times 4H_2O$], and epsomite [$MgSO_4 \times 7H_2O$], were considered.

The rate of CO_2 sequestration via carbonation also is reported in a few recent studies. For example, silicate weathering trapped 102-114 g C m^{-2} y^{-1} within the nesquehonite structure, which corresponded to a two orders-of-magnitude increase over the background rate of CO_2 uptake predicted from arctic and subarctic river catchment data (Wilson et al. 2011). The predicted rate for CO_2 sequestered in ultramafic mine-tailings ranged between 600 and

1000 g C m^{-2} y^{-1}, with the rate being sensitive to CO_2 diffusion through the mineral waste (Bea et al. 2012). The rate of CO_2 diffusion into different soils and under different conditions imposed by climate change is not known. Another study focused on re-vegetated mine spoils over a 19-year period in India. These mine spoils acted as source of pollutants with respect to air dust and heavy metal contamination to soil and water (Tripathi, Singh and Nathanail 2014). However, they also can act as a significant sink for atmospheric CO_2 through combined plant succession and soil formation (an accumulation of total C in total plant biomass, mine soil, and soil microbial biomass of 44.5, 22.9, and 1.8 t ha^{-1}, respectively) (Tripathi, Singh and Nathanail 2014). Total sequestered C increased by 712% after 19 years (a sequestration rate of 364 g C m^{-2} y^{-1}).

There are two additional issues that are important in this discussion. The first issue is related to the necessity to determine the rate-limiting step in a sequence of steps that lead to carbonate formation. Studies have reported that dissolved IC concentrations decreased during dypingite precipitation, indicating that the rate of CO_2 uptake into solution was outpaced by the rate of C fixation via precipitation and implying that the CO_2 gas uptake is the rate-limiting step to CO_2 fixation (Wilson et al. 2010), although other variables should be considered as well. Further studies to elucidate this important aspect of mineral carbonation are required.

The second issue has to do with the possibility of formation of other phases (e.g., sulfates) during tailings weathering, in addition and/or in competition with the carbonates. For example, the relatively unstable Na-Mg hydrated double salt konyaite [$Mg(SO_4)_2 \times 5H_2O$] may be formed in tailings at the Mount Keith mine in Western Australia (Mills et al. 2010). These authors found that Mg bound to sulfate mineral phases reduced the overall potential of mine-tailings to sequester atmospheric C in Mg carbonates, such as hydromagnesite. In addition, amorphous sulfates that formed during konyaite transformation were highly reactive and may contribute to acid mine drainage if present in sufficiently large quantities, and may dissolve carbonate phases that have already sequestered C (Mills et al. 2010), releasing it back to solution and/or atmosphere. The regional and global effects of these processes are currently unknown.

Another recent study illustrated the potential for managing urban soils as tools of C capture and storage, and demonstrated the importance of studying C storage in engineering urban anthropogenic soils (Washbourne, Renforth and Manning 2012). According to the authors, the potential for using engineered urban soils to capture and store atmospheric C was high. In this study, anthropogenic soils, which contained substantial quantities of minerals that are rich in Ca and Mg (21.8 ±4.7% wt $CaCO_3$) derived from demolition activity (particularly cement and concrete), were systematically sampled. Isotopic analysis suggested that up to 39.4 ± 8.8% of the carbonate C has been captured from the atmosphere through hydroxylation of dissolved CO_2 in high-pH solutions, while the remaining carbonate C was derived from lithogenic sources. The CO_2 removal rate was 12.5 kg CO_2 Mg^{-1} y^{-1} (Washbourne, Renforth and Manning 2012).

2.2.4. Inorganic C Flux Out of Soil: Transport to Subsoil/Groundwater, Rivers, Lakes, Oceans

Another important aspect of soil weathering is related to the fate of the weathering reaction products (such as dissolved IC and aqueous species of different chemical elements), which will either precipitate and undergo other in situ processes in soils (such as adsorption or uptake by ecosystems), or be transported via surface and groundwater to rivers, lakes, and oceans (Cao et al. 2012). A substantial amount of the atmospheric C taken up on land through photosynthesis and chemical weathering is transported laterally along the aquatic continuum from upland terrestrial ecosystems to the ocean (Regnier et al. 2013). Rivers transport the soluble products of weathering [e.g., cations, alkalinity, silicic acid (H_4SiO_4), etc.] to the oceans, where they are used by marine ecosystems (Street-Perrott and Barker 2008). Data on river-water quality from 42 monitoring stations in the Lower Mekong Basin obtained during the period 1972–1996 was used to relate solute fluxes with controlling factors such as chemical weathering processes (Li, Lu and Bush 2014). Calcium and bicarbonate dominated the annual ionic composition, accounting for ~70% of the solute load (Li, Lu and Bush 2014). In oceans, biological C pumps (which export OM and $CaCO_3$ to the deep ocean) may be altered by the resulting influx of elements and alkalinity (Hartmann et al. 2013).

Anthropogenic perturbation may have increased the flux of C to inland waters by as much as 1.0 Pg C y^{-1} since pre-industrial times, mainly resulting from enhanced C export from soils (Regnier et al. 2013). However, although it may appear that this process has potential to significantly increase ocean alkalinity, which can partially counteract ocean acidification associated with the current marked increase in atmospheric CO_2, studies have reported that most of the additional C input to upstream rivers was either emitted back to the atmosphere as C dioxide (~0.4 Pg C y^{-1}) or sequestered in sediments (~0.5 Pg C y^{-1}) along the continuum of freshwater bodies, estuaries, and coastal waters, leaving only a perturbation C input of ~0.1 Pg C yr^{-1} to the open ocean (Regnier et al. 2013).

Release of potential contaminants in surface waters as a result of accelerated weathering (Todd et al. 2012) and other climate-change related effects [e.g., the effect of warming temperatures on nitrate concentration of streams (Baron et al. 2009), and the hydrochemical effect of drought during hydrological episodes in streams (Laudon et al. 2004)], and the increase in surface water temperature [e.g., the increase in Columbia River water temperatures in response to climate change (Caldwell et al. 2013)] may significantly or even dramatically change elemental balances; affect microbially mediated reactions and processes occurring in rivers, lakes, and oceans; and/or affect aquatic life [e.g., may contribute to reduced fish survival and lower population productivity (Isaak, Wollrab, et al. 2012; Isaak, Muhlfeld, et al. 2012; Zeigler et al. 2012).

Another recent study illustrated how changes in hydrologic regime may cause changes in biogeochemical processes that exacerbate the danger to aquatic ecosystems (Crouch et al. 2013). According to these authors, a major water quality concern in the Rocky Mountains is acid rock drainage, which causes acidic conditions and high metal concentrations in streams. They found the 30-year water quality record for the upper Snake River in Colorado to show that summer low-flow Zn concentrations have increased four- to six-fold concurrently with increases in mean annual and summer temperatures and a two- to three-week advancement in spring snowmelt. They also found that the main source of Zn and other metal loads to the upper Snake River is a tributary draining an alpine area rich in disseminated pyrite. The authors presented several potential explanations, all related to climate change, which might explain this phenomenon such as 1) decreasing pH in the tributary of interest resulting in mobilization of metals from the wetland and hyporheic zone; 2) the geochemistry of groundwater inflows to the wetland may be changing; and 3) wetland soils may be drying out with longer, warmer summers. Studies are needed to determine the magnitude and fate of various chemical elements released from weathering in different soils and under changing climate conditions.

Another extremely important topic is that climate change will affect groundwater resources and there is a need for quantitative predictions of climate-change effects on groundwater recharge, which may be valuable for effective management of future water resources (Crosbie et al. 2013). The study by Crosbie et al. (2013) reported that predicted changes in recharge between dry and wet future climate scenarios encompassed both an increase and decrease in recharge rates, with the magnitude of this range greater than 50% of the current recharge. Another study developed a new method for identifying triggers of hydrologic droughts by examining the association among various hydro-climatic variables and stream flows (Maity et al. 2013). These and other future studies are important for understanding how crucial water will be in agriculture and food production in a warming world. The impact of climate change on groundwater resources is an important topic that deserves full attention of the scientific community.

2.2.5. Coupled Cycles of C and Other Elements

The C cycle is coupled with the cycles of other elements that are commonly present in terrestrial systems, with Si being one of the elements. There are two important aspects of coupled Si-C cycling related to weathering of silicate minerals. The first aspect has to do with the fact that coupled Si-C cycles are driven by plant action and play a crucial role in the regulation of atmospheric CO_2 (Song et al. 2012). These authors claimed that the processes involved in the coupled cycles of these two elements included plant-enhanced silicate weathering, phytolith formation and solubilization, secondary aluminosilicate accumulation, phytolith occlusion of C, and physico-chemical protection of OC in soils. They also claimed that there was increasing evidence of biological pumping of Si in terrestrial ecosystems, suggesting that complex feedback loops existed among the processes within the coupled Si-C cycles and offering promising new possibilities for enhancing atmospheric CO_2 sequestration. They concluded that organic mulching, rock powder amendment, cultivating Si-accumulating plants, and partial plant harvesting are potential measures that may allow for long-term manipulation and biogeochemical sequestration of atmospheric CO_2 in soil-plant systems (Song et al. 2012). Other studies emphasized the importance of Si-accumulating plants (e.g., grasses, sedges, palms; some temperate deciduous trees and conifers; and many tropical hardwoods), which deposited significant amounts of amorphous hydrated silica in their tissues as opal phytoliths (Street-Perrott and Barker 2008). This paper reviewed the biological role of Si in higher plants, the impact of vegetation on rates of chemical weathering, and the fluxes of Si through catchment ecosystems, lakes, and rivers. The authors stated that on geological time-scales, the global geochemical cycles of C and Si were coupled by the drawdown of atmospheric CO_2 through chemical weathering of Ca- and Mg-silicate minerals in continental rocks (Street-Perrott and Barker 2008), which is the second aspect of coupled Si-C cycling in soils [i.e., release of carbonate-forming cations such as Mg, Fe, and Ca and formation of carbonate minerals (see previous sections of this paper)].

2.3. Soil/Geo-engineering Techniques for C Sequestration

Over the last few years, a series of papers have been published about "enhanced weathering" (which is a term used to describe a soil/geo-engineering concept based on the process of soil-mineral weathering and carbonate precipitation) that is proposed to be used to reduce rising CO_2 levels by spreading fine-powdered olivine on farmland or forestland (Schuiling and Krijgsman 2006; Köhler, Hartmann and WolfGladrow 2010; Köhler et al. 2011; Schuiling 2012; Schuiling et al. 2011; Olsson et al. 2012; Zevenhoven et al. 2011; King et al. 2010; Hartmann et al. 2013; ten Berge et al. 2012).

Enhanced weathering may sequester a significant amount of CO_2 in soils at relatively fast time-scales. Progress in this area of research has been encouraged by the slow deployment of large-scale underground storage of CO_2 and the fact that large amounts of suitable, relatively unstable, minerals and rocks are available worldwide (Zevenhoven, Fagerlund and Songok 2011). In addition, the method offers leakage-free CO_2 fixation that does not require post-storage monitoring (Zevenhoven, Fagerlund and Songok 2011). However, published studies on how efficiently this geo/soil-engineering approach would work in different soils and under different climate conditions are scarce.

Enhanced weathering, like natural weathering, involves the dissolution of Ca- and Mg-bearing silicate minerals and the subsequent release into the aqueous phase of Ca and Mg where they may react with dissolved CO_2 in the form of bicarbonate (HCO_3^-) and carbonate (CO_3^{2-}) ions, to form carbonate minerals such as calcite ($CaCO_3$), magnesite ($MgCO_3$), and dolomite [$CaMg(CO_3)_2$] (Renforth 2012). Weathering and subsequent precipitation of Ca- and Mg-bearing carbonates are the main processes that contribute to CO_2 gas removal from the atmosphere (Schuiling and Krijgsman 2006). Artificially enhanced silicate weathering would not only operate against rising temperatures but would, indirectly, reduce ocean acidification, because it influences the global climate via the IC cycle (Köhler, Hartmann and Wolf-Gladrow 2010). These authors reported that the potential sequestration was up to 1 Pg C y^{-1} if olivine would be distributed as fine powder in the humid tropic regions of the Amazon and Congo River catchments, but they also emphasized that this causes pH values in the rivers to rise to 8.2, and the dissolution rate was limited by the saturation concentration of H4SiO4. Other researchers claimed that the presence of a new Mg-silicate phase and the formation of secondary products at the olivine surface may limit the extent of olivine to carbonate conversion and understanding the relationship between the formation of carbonate and other phases is important for predicting the impact of possible passivating layers on the extent and rate of reaction (King, Plumper and Putnis 2010).

In another paper (Schuiling, Wilson and Power 2011), the authors disputed the conclusion by (Köhler, Hartmann and Wolf-Gladrow 2010) that the saturation of H4SiO4 would limit olivine dissolution. Schuiling et al. (2011) claimed that the H4SiO4 may be consumed abiotically [i.e., precipitation of opaline silica, quartz and/or formation of phyllosilicate clay minerals] and/or biotically [i.e., by vascular plants and diatoms in lakes and rivers] [(Schuiling, Wilson and Power 2011) and references therein]. Based on these arguments, Schuiling et al. 2011 claimed that the saturation concentration of H4SiO4 would be never reached during natural weathering of pulverized ultramafic rock. This is definitely an open discussion and further research efforts are needed. Additional information on the enhanced weathering geo-engineering approach can be found in other papers (Schuiling 2012; Schuiling and Krijgsman 2006; Renforth 2012; Hartmann et al. 2013; Moosdorf et al. 2014; Kojima et al. 1997; Lackner et al. 1995; Lackner et al. 1997; Power et al. 2013; Krevor and Lackner 2009; Ruiz-Agudo et al. 2013; Sissmann et al. 2014).

Another method to sequester atmospheric CO_2 via artificially accelerated weathering involves the following two steps (Kakizawa et al. 2001):

1) $CaSiO_3 + 2CH_3COOH \rightarrow Ca^{2+} + 2CH_3COO^- + H_2O + SiO_2$

2) $Ca^{2+} + 2CH_3COO^- + CO_2 + H_2O \rightarrow CaCO_3\downarrow + 2CH_3COOH$

According to these authors, the first step was the extraction of Ca ions by acetic acid from Ca silicate (e.g., wollastonite). The second step was Ca carbonate precipitation promoted by the presence of CO_2. Other recent studies have evaluated how alkaline silicate mineral-based CO_2 sequestration can be achieved using environmental biotechnological processes that involve an acid-producing reaction to enhance silicate dissolution and a subsequent alkalinity-producing step to precipitate carbonates (Salek et al. 2013).

In concluding this discussion about enhanced weathering of silicate minerals as a method for sequestering atmospheric CO_2 concentration, one should emphasize that many issues related to weathering rates in different soils and conditions and impacts to soils and water resources are not fully understood, and further investigation in this area is required to reduce the many uncertainties associated with this method.

3.0. CLIMATE-CHANGE INDUCED SOM TRANSFORMATION AND MINERALIZATION

3.1. SOM in Soils

The organic matter content of soils is one of the most important soil properties that define soil fertility and, to a large extent, control the chemical and biological behavior of the soil system. For this reason, different aspects of SOM formation, transformation, mineralization, and interaction with soil minerals or other organic molecules have attracted the attention of many soil scientists over the last decades. However, although this is a relatively well explored area of research, SOM studies at the molecular, nano, and micron scales have intensified especially in recent years because of technological advances and the relevance to climate change, which is expected to affect in unprecedented ways the life on earth.

Organic C incorporated into SOM may play a major role in controlling soil behavior as a sink or source for atmospheric CO_2 (Ghee et al. 2013), thus contributing significantly to the global C cycle. Because of predicted climate change and the need for a more unified approach to mitigate greenhouse gas emissions, the soil's ability to sequester and/or release C into atmosphere, and thus to act as a C sink or source for atmospheric CO_2, has received growing attention (Solomon et al. 2012). For example, enhanced soil respiration in response to global warming may substantially increase atmospheric CO_2 concentrations above the anthropogenic contribution, depending on the mechanisms underlying the temperature sensitivity of soil respiration (Tucker et al. 2013). The uncertainty associated with how projected climate change will affect global C cycling could have a large impact on predictions of soil C stocks (Haddix et al. 2011) and the long-term changes in soil OC storage (Conant et al. 2011).

A new set of scientific questions have surfaced recently about the soil sink/source behavior and the effects of climate-change variables on the positive or negative CO_2 gas fluxes between terrestrial systems and atmosphere, and other important soil processes and reactions. In addition, climate-change variables may induce significant or even dramatic shifts in SOM budgets and distribution among different pools, with short- and long-term effects that are currently unknown. To address these and other related questions and uncertainties, an increasing number of scientists from different disciplines and communities are using state-of-the-art microscopic and spectroscopic techniques (Remusat et al. 2012; Kleber et al. 2011; Vogel et al. 2014; Kaiser and Guggenberger 2007; Solomon et al. 2012; Kinyangi et al. 2006; Solomon et al. 2007; Lehmann et al. 2008; Erhagen et al. 2013) with the objective of improving our understanding of different aspects of SOM transformations and interactions within soil matrices. To address emerging issues in these research areas, scientists also are using a variety of research instruments that would allow interrogations of soils and SOM under transient dynamic conditions [e.g., dynamic transmission electron microscopy (DTEM) used in studies conducted at the Environmental Molecular Science Laboratory in Richland, Washington); instruments that can probe SOM heterogeneities at incredibly small spatial resolutions (Milne et al. 2011; Lehmann et al. 2008) and extremely fast temporal scales (involving DTEM); and sophisticated geostatistical models such as those used to study spatial distribution of soil organic carbon (SOC) at large scales (Martin et al. 2014), or other models such as the Dual Arrhenius and Michaelis-Menten kinetics models (Davidson et al. 2012; Paterson and Sim 2013).

This section is not intended to be a comprehensive review, but instead presents a focused effort on a variety of recent developments and an attempt to uncover emerging and/or unresolved research gaps on the effects of climate-change variables on SOM budgets and pools, protection, mineralization, and temperature sensitivity. The reader is referred to a number of recent review papers for an in-depth coverage and discussion on 1) the evolution of concepts of the chemical and physical nature of SOM and arguments that have led to increased skepticism about the humification concept and usefulness of operationally defined "humic substances" (Kleber and Johnson 2010); 2) the amount of C stored in soils globally, the potential for C sequestration in soil, and successful methods and models used to determine and estimate C pools and fluxes (Stockmann et al. 2013); 3) SOM characterization, isolation and fractionation, pools, and formation of recalcitrant components of SOM (Maia et al. 2013); 4) earth science models that should simulate microbial physiology to more accurately project climate-change feedbacks (Wieder et al. 2013); 5) methods used to characterize SOM structure, source, and degradation that have enabled precise observations of SOM and associated ecological shifts

(Simpson and Simpson 2012); 6) some important and challenging aspects of soil extracellular enzyme research (Burns et al. 2013); 7) SOC sequestration under different tillage systems in China (Zhang et al. 2014); 8) the pool of thaw-vulnerable OC of the permafrost based on field data and extrapolation using geospatial data (Strauss et al. 2013); and 9) impact of permafrost thaw on OM chemistry which could intensify the predicted climate feedbacks of increasing temperatures, permafrost C mobilization, and hydrologic changes (Hodgkins et al. 2014).

3.1.1. SOM Budget and Pools

Globally, a significant C amount is stored in the world's soils, including peatlands, wetlands, and permafrost (Davidson and Janssens 2006; von Lutzow and Kogel-Knabner 2009; van Groenigen et al. 2014). On a global scale, soils contain approximately 2344 Pg C in the top 3 m of soil, with about 1502 Pg in the first meter, and 491 and 351 Pg C for the second and third meters, respectively (Jobbagy and Jackson 2000). In addition, SOM contains more than three times as much C as either the contemporary atmosphere or terrestrial vegetation (Schmidt et al. 2011; Erhagen et al. 2013); therefore, even small changes to its inventory may have major implications for atmospheric CO_2 concentrations (Erhagen et al. 2013; Stockmann et al. 2013; van Groenigen et al. 2014). Clearly, soils may play a key role in modulating climate change (van Groenigen et al. 2014). For this reason, our ability to predict and ameliorate the consequences of global warming depends in part on a better understanding of the distributions and controls over SOM (Jobbagy and Jackson 2000).

The SOM originates from different sources such as residues from plant, microbes, or other living organism in soils and, chemically, is a complex mixture of many organic compounds. Even when it originates from the same source, the chemical composition of the SOM is remarkably diverse. For example, plant residues often are complex mixtures of mainly polysaccharides [i.e., starch, cellulose, hemicellulose and pectin (50–60%); lignin (15–20%), polyphenols (e.g., tannins), chlorophyll, cutin and suberin, lipids and waxes (10–20%)] (von Lutzow et al. 2006). These organic compounds may have different fates and follow different transformation pathways, further contributing to SOM chemical composition diversity. A critical examination of published data obtained from many molecular-scale studies revealed that humic substances are diverse with components spatially segregated at the nanometer scale (Sutton and Sposito 2005). Other studies have found that humic substances, which represent the largest pool of recalcitrant SOM in the terrestrial environment, are a very complex mixture of microbial and plant biopolymers and their degradation products that do not represent a distinct chemical category (Kelleher and Simpson 2006). This is the main reason why SOM pools are difficult to define based on the chemistry and/or molecular identity of the organic compounds that are present in these pools.

Nevertheless, for modeling and investigative purposes, SOM pools are operationally defined as 1) a labile pool (short residence time) and 2) a recalcitrant (refractory) pool (long residence time), which includes humin, and fulvic and humic acids. Based on the stabilization and protection mechanisms, SOM pools are defined as 1) unprotected; 2) physically protected from decomposition via micro-aggregation; 3) stabilized via intimate association with soil fine fractions (silt and clay); and 4) biochemically stabilized through the formation of recalcitrant SOM compounds; each of these pools behaves differently in terms of extent and rate of decomposition (Six et al. 2002), but their individual and combined contributions to the CO_2 gas flux coming out of soils, remains unclear.

As the largest pool of terrestrial OC, soils interact strongly with atmospheric composition, climate, and land cover changes (Jobbagy and Jackson 2000). If C stored belowground is transferred to the atmosphere by a warming-induced acceleration of its decomposition, a positive feedback to climate change would occur (Davidson and Janssens 2006). Conversely, if increases of plant-derived C inputs to soils exceed increases in decomposition, the feedback to climate change would be negative (Davidson and Janssens 2006). The fluxes of SOC vary in response to a host of potential environmental and anthropogenic driving factors and important questions have been raised, such as (Stockmann et al. 2013):

- What is the average net change in SOC due to environmental conditions or management practices?
- How can SOC sequestration be enhanced to achieve atmospheric CO_2 mitigation?
- Will this secure soil quality and preserve soil resources so that they will provide sufficient food and fiber to a growing world population?

The first and the second questions were addressed in a study conducted at Rothamsted (United Kingdom) (Fornara et al. 2011). In that study, the researchers investigated the combined effect of liming on microbial respiration and net OC accumulation using experimental data spanning 129 years. They found that liming contributes to higher rates of respiration from soil (thereby potentially reducing soils ability to act as a CO_2 sink; however, it stimulated biological activity that, despite increasing soil respiration rates, led to plant C inputs being processed and incorporated into resistant soil organo-mineral pools, shielding SOM from weathering. The net OC sequestration reported in this study, measured in the 0–23 cm soil layer at different time intervals since 1876, was 2–20 times greater in limed than in unlimed soils). The authors concluded by stating that the greater OC sequestration in limed soils strongly reduced the global warming potential of long-term liming to permanent grassland suggesting the net contribution of agricultural liming to global warming could be lower than previously estimated (Fornara et al. 2011). Their study demonstrated that the common agricultural practice of applying lime might be an effective mitigation strategy especially because it can be associated with a reduced use of nitrogen (N) fertilizers, which are a key cause of increased greenhouse gas emissions from agro-ecosystems (Fornara et al. 2011). This example clearly illustrates that changes in soil inorganic chemistry may have significant impacts on OC storage and its distribution or redistribution within pools in which it has become less bioavailable.

A few studies have addressed issues related to the SOM budget under climate-change conditions (i.e., atmospheric CO_2 enrichment, intensive precipitation or prolonged drought, higher temperatures and heat waves) and different soil management approaches. One study was conducted with soils from the U.S. Great Plains, which contained enormous stocks of SOC and soil organic nitrogen (SON) that would be vulnerable to predicted climate and land use changes (Follett et al. 2012). The authors found that soil C and N stocks under different land uses were equally vulnerable to increased temperature and decreasing water availability. They reported that an increase in mean annual temperature of 1°C (1.8°F) could cause a loss of 486 Tg SOC and a loss of 180 kg SON ha^{-1} from the top 10 cm of soil over 30 years. They concluded that combined increased temperature and conversion from the conservation program to cropland could decrease the existing SOC pool sink, but improved soil management and increased water availability may help offset these losses in the U.S. Great Plains (Follett et al. 2012). Most C stocks in soils are comprised of SOM with turnover times of decades to centuries, and understanding the response of these C pools to climate change is essential for predicting longer-term changes in soil C storage (Conant et al. 2011). However, additional and much broader efforts are required to fully address the question posed above.

The SOM turnover times may be accelerated because of increasing CO_2 concentration in the atmosphere as it was demonstrated in a study that investigated the effect of rising atmospheric CO_2 on combined responses of stimulated plant growth (which result in soil C addition) and microbial decomposition (which result in soil C removal) (van Groenigen et al. 2014). The results showed that atmospheric CO_2 enrichment stimulates both the input of C in soil (+19.8%) and the turnover of C in soil (+16.5%) (van Groenigen et al. 2014). They also reported that the increase in soil C turnover with rising CO_2 led to lower equilibrium soil C stocks than expected from the increase in soil C input alone, indicating that the net outcome of these combined effects was negative and results in a general mechanism that limits C accumulation in soil (van Groenigen et al. 2014). This pioneering study should be followed by many more so other important aspects of soil and ecosystem response to climate change will be elucidated.

3.1.2. SOM Saturation Limit

One important concept related to the soil C budget concerns soil finite capacity or soil OC saturation limit and SOM dynamics within the framework initially described by Six et al. (Six et al. 2002) and explored further in other studies (Stewart et al. 2007; Stewart, Plante, et al. 2008; Stewart, Paustian, et al. 2008; Qin et al. 2013; Feng et al. 2014). Soil OC saturation behavior predicts that soil OC storage efficiency observed under field conditions decreases as the soil approaches OC saturation (Six et al. 2002). These authors suggested that physicochemical characteristics inherent to soils usually define the maximum protective capacity for SOM, and limit increases in SOM (i.e., C sequestration) with increased organic residue inputs (Six et al. 2002). Other authors have noticed that current estimates of soil OC storage potential are based on models or factors that assume linearity between OC input levels and OC stocks at steady-state, implying that SOC stocks could increase without limit as C inputs increase (Stewart et al. 2007). In their long-term field experiments, these authors found that some soils showed little or no increase in steady-state SOC stock with increasing C inputs, suggesting saturation. In this and other studies, the authors concluded that the saturation of soil OC did occur, and therefore, the greatest efficiency in soil C sequestration will be in soils further from OC saturation (Stewart et al. 2007), or in soils with low OC contents and in degraded lands (Stewart, Paustian, et al. 2008).

In another study (Stewart, Plante, et al. 2008), four different soil OC pools [e.g., unprotected (free particulate organic matter); physically protected (microaggregate-associated C); chemically protected (silt- and clay-associated C), and biochemically protected (non-hydrolyzable C)] were considered. These authors assessed two contrasting models of SOC accumulation, one with no saturation limit (i.e., linear first-order model) and one with an explicit soil OC saturation limit (i.e., C saturation model). They found that the two sites with the greatest SOC range showed OC saturation behavior in the chemically, biochemically, and some mineral-associated pools of the physically protected pool. Importantly, the unprotected pool and the aggregate-protected showed linear, non-saturating behavior.

Evidence of C saturation of chemically and biochemically protected SOC pools was observed at sites far from their theoretical OC saturation level, while saturation of aggregate-protected pools occurred in soils closer to their OC saturation level (Stewart, Plante, et al. 2008). In a more recent study, data from worldwide, long-term agricultural experiments were used to develop two statistical models to determine the saturated SOC level in upland and paddy agro-ecosystems (Qin, Huang and Zhuang 2013). These authors found that Chinese croplands had relatively low SOC contents in comparison to the global average and great potential for C sequestration under improved agricultural management strategies; the time required to reach SOC saturation in Chinese cropland was highly dependent on the management practices applied (Qin, Huang and Zhuang 2013).

Based on the assumption that soil OC saturation behavior predicts that soil OC storage efficiency observed under field conditions decreases as a soil approaches C saturation, other researchers have hypothesized that this may be due to a decline in SOM stability as the result of changes in the type, strength, or turnover time of organo-mineral interactions with increasing organic C input (Feng et al. 2014). However, results from a series of batch sorption experiments did not support the hypothesis that SOM stability decreases with increasing C loading.

The concept of C saturation is called into question in a recent study by Vogel et al. (Vogel et al. 2014). The results from this study provided evidence that only a limited proportion of the clay-sized surfaces contributed to OM sequestration, providing a different view of C sequestration in soils and the widely used C saturation estimates (Vogel et al. 2014).

3.1.3. SOM Accumulation in Subsoil

Conceptual models developed based on results collected from many studies conducted with soils from different regions consider that the stability of SOM depends, among other factors, on location within the soil profile. Studies have confirmed that the subsoil OC pool is significant. The percentage of SOC in the top 20 cm (relative to the first meter) averaged 33%, 42%, and 50% for shrublands, grasslands, and forests, respectively (Jobbagy and

Jackson 2000). In shrublands, the amount of SOC in the second and third meters was 77% of the amount found in the first meter; in forests and grasslands, the totals were 56% and 43%, respectively (Jobbagy and Jackson 2000). These results clearly demonstrate the tendency of the SOC to accumulate (i.e., stabilize) in the subsoil.

Results from other studies have shown that the relevance of spatial inaccessibility and organo-mineral interactions for SOM stabilization increased especially in subsoil (von Lutzow et al. 2006). However, the reasons for enhanced OC stabilization in subsoil horizons are currently poorly understood (Chabbi et al. 2009), and whether the stabilization mechanisms are different in subsoil than topsoil remains unresolved. Some other important scientific issues related to SOM accumulation in subsurface, C cycling and climate change are identified and discussed below:

1) *How is SOM transported from topsoil to subsoil?* Ideas on how SOM is transported from topsoil to subsoil, in addition to the most common mode of transportation (i.e., DOM moving down the soil profile with advective flow), were presented in a few recent studies. For example, one study found that long-term C storage in short-range ordered minerals occurred via chemical retention with dissolved aromatic acids derived from plant litter, which were carried along preferential flow-paths to deeper B horizons (Kramer et al. 2012). Another study, conducted with basaltic soils from Hawaii exposed to high rainfall and anoxic conditions that facilitated Fe(III) (oxyhydr)oxide reductive dissolution (i.e., climate-controlled redox dynamic condition), showed that colloidal dispersion during Fe-reducing conditions mobilized high concentrations of C from the pool of C strongly associated with the Fe(III) phases, which may then migrate to deep mineral horizons (Buettner et al. 2014).

2) *How can SOM spatial heterogeneity and the effects of hydrology be characterized and represented in models and how would this vary as a function of the scale of the models?* The effect of hydrology and heterogeneous distribution on SOM preservation in subsoil was investigated in a paper by Chabbi et al., 2009. In the subsoil of three agricultural lands, compartments of visually identifiable zones in the form of tongues (with modern age SOM) and the adjacent soil matrix (with several thousand-year-old SOM) were present in deep soil horizons (60-140 cm depth) (Chabbi, Kogel-Knabner and Rumpel 2009). The results indicated that tongues had two times higher C

content and were depleted in N with regards to the adjacent soil matrix. Twenty percent more hydrofluoric-acid soluble C was in the soil matrix compared to tongues. The authors of this study suggested that, in tongues, fresh C input by preferential flow and/or roots led to higher SOM turnover compared to the soil matrix. The effect of hydrology and heterogeneous distribution of stabilized SOM within the subsoil matrix must be taken into account when studying C sequestration in deep soil horizons (Chabbi, Kogel-Knabner and Rumpel 2009).

3) *What is the SOM stabilization mechanism in subsurface?* Studies about SOM stabilization mechanisms have shown that OC stability in deep soil layers depends on the supply of fresh plant-derived C to the subsoil, which stimulated the microbial mineralization of 2567 ±226-year-old C (Fontaine et al. 2007). These results supported the idea that, in the absence of fresh OC (an essential source of energy for soil microbes), the stability of OC in deep soil layers was maintained (Fontaine et al. 2007). Other studies have also shown that SOM mineralization in many soils generally increased after addition of carbohydrates, amino acids, or simple organic acids, thus indicating that stability may be caused by substrate limitations (Marschner et al. 2008). Under climate-change conditions, a lack of supply of fresh OC may prevent the decomposition of the SOM in deep soil layers in response to temperature changes. On the other hand, intensive rainfall and soil management practices, such as changes in land use and application of different agricultural practices that increase the distribution of fresh C along the soil profile, may stimulate the loss of ancient buried C (Fontaine et al. 2007).

4) *What is the role of subsurface naturally reduced zones (NRZ) or other organic-rich systems on C and elemental cycling?* NRZs represent a SOM pool common in many environments. Subsurface NRZs are characterized by transient anaerobic and aerobic conditions and are places where the biogeochemical cycles of C and other elements (such as Fe) or contaminants (such as U) are strongly interlinked and overlap in remarkable and rather complex ways (Qafoku et al. 2009; Campbell et al. 2012; Qafoku et al. 2014). There are examples in the literature showing how cycling of other elements affect the C cycle other organic-rich systems (e.g., humid tropical forests), similar to NRZs. A recent study found that microbial Fe reduction generated reduced Fe^{2+} under anaerobic conditions, which oxidized to Fe^{3+}

under subsequent aerobic conditions (Hall and Silver 2013). These authors demonstrated that Fe^{2+} oxidation stimulated SOM decomposition via two mechanisms: 1) OM oxidation [via generation of reactive O2 species during Fe^{2+} oxidation] and 2) increased DOC availability [via proton generation during Fe^{2+} oxidation] (Hall and Silver 2013). This study concluded that Fe oxidation coupled to SOM decomposition contributed to rapid rates of C cycling across humid tropical forests in spite of periodic O_2 limitation, and may help explain the rapid turnover of complex C molecules in these soils (Hall and Silver 2013). Coupling of C and Fe cycles also is common in marine sediments as well, where the associations between OC and Fe formed primarily through co-precipitation and/or direct chelation promote the preservation of OC in sediments of various mineralogy from a wide range of depositional marine environments (Lalonde et al. 2012).

5) *What is the sensitivity of subsoil SOM to global change drivers?* The sensitivity of SOC at different soil depths to global change drivers is another topic receiving increasing attention because of its importance in the global C cycle and its potential feedback to climate change (Albaladejo et al. 2013). According to these authors, the relative importance of climatic factors decreased with increasing depth, and soil texture became more important in controlling SOC in all land uses (Albaladejo et al. 2013). Because of climate change, impacts will be much greater in surface SOC, so the strategies for C sequestration should be focused on subsoil sequestration, which was hindered in forestland due to bedrock limitations to soil depth. In these conditions, sequestration in cropland through appropriate management practices was recommended (Albaladejo et al. 2013). A better knowledge of the vertical distribution of SOC and its controlling factors will help scientists predict the consequences of global change (Albaladejo et al. 2013).

Other factors that may control SOM distribution with depth and its preservation are vegetation, soil types, parent material, and land use (Schrumpf et al. 2013); no-tillage management as a practice capable of offsetting greenhouse gas emissions and its ability to sequester C in soils combined with improved N management techniques (Six et al. 2004); and changes in tropical land use and cultivation to control SOM status in subsoil because cultivation modifies the distribution of the more labile fractions of the SOM (Guimaraes et al. 2013).

3.2. SOM Protection Mechanisms

Mechanisms and pathways of SOM stabilization have received increasing attention recently because of their relevance in the global C cycle and global warming. The following discussion will be focused on these and other stabilization mechanisms, and a variety of scientific issues related to SOM stabilization and mineralization in soils under conditions imposed by the climate change.

Numerous attempts have been made to understand SOM protection mechanisms that operate in different soils and under different conditions (Six et al. 2002; von Lutzow et al. 2006; Schrumpf et al. 2013). The predominant SOM protection mechanisms in temperate soils, discussed in a comprehensive publications by von Lutzow et al. (von Lutzow et al. 2006) are:

1) selective preservation due to structural composition including plant litter, rhizodeposits, microbial products, humic polymers, and charred OM;
2) spatial inaccessibility of SOM against decomposer organisms due to occlusion, intercalation, hydrophobicity and encapsulation; and
3) stabilization by interaction with mineral surfaces (Fe, Al, Mn oxides, and phyllosilicates) and metal ions (von Lutzow et al. 2006). The protection mechanisms may also operate simultaneously at different stages of SOM decomposition (von Lutzow et al. 2006), and may be a function of site-specific, spatial- and time-dependent dynamic conditions that often are combined with complex scale-dependent effects.

3.2.1. Selective Preservation

When it comes to the stabilization mechanisms of SOM as a complex mixture of identifiable biopolymers (Kelleher and Simpson 2006) rather than a chemically complex material (such as humic and/or fulvic acids), one of the predominant arguments for many years has been about the resistance of SOM to mineralization, and whether this apparently inherent chemical or biochemical property makes some SOCs more resistant than others so that they are selectively preserved. The paradigm of what is called "intrinsic recalcitrance" is therefore based on the idea that some naturally occurring organic molecules in soils can resist decomposition because of thermodynamic stability and/or their unique and resistant molecular structure.

Traditionally, the selective preservation of certain recalcitrant organic compounds and the formation of recalcitrant humic substances have been regarded as an important mechanism for SOM stabilization (Marschner et al. 2008). Solid-state ^{13}C nuclear magnetic resonance studies suggested that the most persistent mineral-bound C was composed of partially oxidized aromatic compounds with strong chemical resemblance to DOM derived from plant litter (Kramer et al. 2012). In addition, soil C turnover models generally divided SOM into pools with varying intrinsic decomposition rates based on the assumption that the chemical structure had primary control over decomposition (Kleber et al. 2011). One should emphasize, however, that determining recalcitrance through experiments is difficult because the persistence of certain SOM pools or specific compounds also may be the result of other stabilization mechanisms, such as physical protection or chemical interactions with mineral surfaces (Marschner et al. 2008).

Some authors have recently disputed the view that SOM stabilization is dominated by the selective preservation of recalcitrant organic components that accumulate in proportion to their chemical properties (von Lutzow et al. 2006). In contrast, they have shown that the soil biotic community is able to disintegrate any organic compound, making the molecular recalcitrance of SOM relative rather than absolute (von Lutzow et al. 2006). In a recent paper by Kleber (Kleber 2010), the author argued that "... recalcitrance is an indeterminate abstraction whose semantic vagueness encumbers research on terrestrial C cycling." The author proposed a different view to the perceived "inherent resistance" to decomposition not as a material property, but as a logistical problem constrained by 1) microbial ecology, 2) enzyme kinetics, 3) environmental drivers, and 4) matrix protection (Kleber 2010). A consequence of this view would be that the frequently observed temperature sensitivity of SOM decomposition will result from factors other than intrinsic molecular recalcitrance (Kleber 2010; Kleber and Johnson 2010).

In a follow-up study, near edge X-ray absorption fine structure (NEXAFS) spectroscopy was used in combination with differential scanning calorimetry (DSC) and alkaline cupric oxide (CuO) oxidation to test the hypothesis whether the chemical structure has primary control over decomposition (Kleber et al. 2011). This study found that the SOM of an Inceptisol, with a ^{14}C age of 680 years, had the largest proportion of easily metabolizable organic molecules with low thermodynamic stability, whereas the SOM of the much younger Oxisol (107 years) had the highest proportion of supposedly stable organic structures considered more difficult to metabolize. The authors suggested that soil C models would benefit from viewing turnover rate as co-determined by the interaction between substrates, microbial actors, and abiotic driving variables.

3.2.2. Spatial Inaccessibility

The second SOM protection mechanism has to do with the fact that the SOM can be spatially inaccessible and, as a result, unavailable to decomposer organisms due to occlusion, intercalation, hydrophobicity, and encapsulation (von Lutzow et al. 2006).

Occluded SOM is spatially protected due to reduced access for microorganisms and enzymes, reduced diffusion of enzymes within the intra-aggregate space, and reduced diffusion of O2 responsible for the aerobic decomposition of the SOM (von Lutzow et al. 2006). The inter- and intra-aggregate pore-size distribution and SOM occurrence location (either close to the entry of the pores or deep in the remote sites) within aggregates plays an important role in the physical protection of the SOM. However, the size of the pores seems to play an important protective role because the formation of multiple complex bonds per molecule is possible and favored in small pores (Kaiser and Guggenberger 2007). A study conducted with microporous goethite demonstrated that the SOM tightly bound via multiple complex bonds, most likely at the entry of small pores, were resistant to desorption and attack of chemical reagents and probably enzymes (Kaiser and Guggenberger 2007). Another study demonstrated the role of spatial connectivity and pore-size distribution within the intra-aggregate space in affecting the ability of macro-aggregates to physically protect C (Ananyeva et al. 2013). **Intercalation** has to do with SOM sorption into interlayer space of expandable phyllosilicates, a protection mechanism that is operational especially in acidic soils (von Lutzow et al. 2006). However, Eusterhues et al. (2003) found no evidence of intercalated SOM into interlayer spaces of phyllosilicates in two German acid forest soils. Currently, a method has yet to be developed to probe the interlayer space of expandable phyllosilicates for intercalated SOM (Leifeld and Kögel-Knabner 2001). **Hydrophobicity** is another important factor to be considered as one of the SOM protection mechanisms because it may result in decreasing SOM decomposition rates and enhanced aggregate stability [(von Lutzow et al. 2006) and references therein]. Finally, another SOM protection mechanism is the one that involves labile organic matter **encapsulated** inside recalcitrant compounds although evidence of the occurrence of encapsulation is limited [(von Lutzow et al. 2006) and references therein]. It is challenging, however, to prove that this mechanism occurs in soils and advanced research techniques definitely need to be employed for these studies.

3.2.3. Interaction with Minerals

A third protection mechanism is that of SOM stabilization via interactions with mineral surfaces (such as Fe, Al, Mn oxides, and phyllosilicates) and metal ions (von Lutzow et al. 2006). Adsorption of SOM to fine soil particles, most frequently clay-sized particles [or even terrestrial nano-particles (Qafoku 2010)], is a well-known phenomenon occurring commonly in soils. The organo-mineral assemblages are even considered as a separate pool in mineral horizons of forest soils (Gruneberg et al. 2013). In some instances, the SOM consists of a heterogeneous mixture of compounds that display a range of amphiphilic or surfactant-like properties, and are capable of self-organization in aqueous solution (Kleber et al. 2007); however, the adsorption properties of these organic mixtures to soil minerals have not been well studied.

The fundamental aspects of SOM sorption to soil minerals, stabilization, and preservation are demonstrated by a substantial decrease in biological degradability after SOM is sorbed to mineral surfaces (Kaiser and Guggenberger 2000; Zech et al. 1997; Kaiser and Guggenberger 2003; Kaiser et al. 2007; Lopez-Sangil and Rovira 2013). Strong correlations between Fe oxides and SOMs have implied the importance of the former in stabilizing the latter (Wagai and Mayer 2007). Other studies have investigated SOM sorption to soil minerals such as Al and Fe oxyhydroxides (Kaiser and Guggenberger 2000); amorphous $Al(OH)_3$; gibbsite, ferrihydrite, goethite, hematite, and phyllosilicates (kaolinite, illite) (Kaiser and Guggenberger 2003); and ferrihydrite and goethite (Kaiser, Mikutta and Guggenberger 2007). Studies have also looked at binding of lignin from three litters (blue oak, foothill pine, and annual grasses) to five minerals (ferrihydrite, goethite, kaolinite, illite, montmorillonite) (Hernes et al. 2013); and protein, lipid, carbohydrate, oxidized lignin, and carboxyl/carbonyl content interaction with short-range order minerals in soils from Hawaii (Kramer et al. 2012). Finally, studies on fractionation procedures capable of assessing the strength through which mineral-associated SOM is stabilized (Lopez-Sangil and Rovira 2013) may be useful in elucidating SOM interactions and affinity for different minerals in soils.

Because the SOM content in soils is usually positively related to the reactive surface area (Kaiser and Guggenberger 2003), quite often the soil clay content or reactive surface area has been used to estimate the SOM sequestration potential of soils (Vogel et al. 2014). However, only a relatively small portion (i.e., less than 19%) of the clay-sized surfaces of the topsoil binds SOM (Vogel et al. 2014). This study also showed that the new SOM added into the system was preferentially attached to already present organo-mineral clusters with rough surfaces (Vogel et al. 2014) and did not bind to other, apparently available mineral surfaces, indicating that surface properties were more important than the reactive surface area or soil texture, a concept that also was discussed in earlier studies (Kaiser and Guggenberger 2000; Plante et al. 2006).

The increase in SOM stability after interacting with soil minerals was shown in a series of incubation experiments conducted with SOMs of different origins sorbed to a subsoil material (Kalbitz et al. 2005). They found that the fraction of sorbed OC mineralized was much less than the fraction of the same identity in solution (in the absence of solids). They also estimated that the mean residence time of the most stable SOM was increased from 28 years in solution to 91 years after sorption. Different hypotheses have been tested in recent studies with the ultimate goal to determine the mechanistic aspects of SOM protection via interactions with soil minerals. The three-way correlation among SOM concentrations, specific surface areas, and small mesopores observed for many soils and sediments led researchers to develop the hypothesis that enclosure within the relatively small pores might explain the apparent protection of SOM by minerals (Mayer et al. 2004). They tested this hypothesis by examining whether the bulk of SOM resides within small mesopores. They found that, although smaller mesopores have sufficient volumes to contain significant fractions of the total OM, only small fractions of OM reside in them. They also found that OM was preferentially associated with aluminous clay particle edges rather than the largely siliceous clay faces that contribute most surface area and form pore walls. They concluded that, while simple enclosure within smaller mesopores cannot explain protection, network effects working at larger size scales may account for exclusion of digestive agents, resulting in OM protection (Mayer et al. 2004).

In another study (Wagai and Mayer 2007), the authors tested the hypothesis that sorption was important in the stabilization of SOMs by reductively dissolving Fe oxides in a wide variety of soils and measuring OC that was subsequently released. They found that the resultant pool, reductively soluble OC, made up a minor amount of total soil OC in all but one of these soils, indicating that simple sorption reactions do not stabilize the bulk of soil OC in most mineral soils and that SOM stabilization may occur via other mechanisms, such as organo-Fe oxide precipitates (see below) or ternary associations among Fe oxides, the SOM, and other minerals. However, the scientific question regarding why and how OM matter is protected from decomposition when it is associated with minerals in soils still has not been answered.

Sorptive stabilization as a function of SOM chemical composition is another important topic. The extent of sorption of recalcitrant compounds was much larger than sorption of labile compounds (Kalbitz et al. 2005). These authors claimed that stabilization of OM by sorption depended on the intrinsic stability of organic compounds sorbed and that the main stabilization processes were selective sorption of intrinsically stable compounds and strong chemical bonds to the mineral soil and/or a physical inaccessibility of OM to microorganisms (Kalbitz et al. 2005). Other researchers claimed that because sorption of the more labile polysaccharide-derived DOM on mineral surfaces is weaker, adsorptive and desorptive processes strongly favored the accumulation of the more recalcitrant lignin-derived SOM in soils (Kaiser and Guggenberger 2000). Additional research is definitely needed to address the issue of increases in stability due to sorption and to show abiotic control on the mineralization rate of the sorbed SOM.

Types of binding mechanisms of SOM with minerals are another important issue that deserves more attention. Sorption of DOM derived from the oxidative decomposition of lignocellulose to Al and Fe oxyhydroxides involves strong complexation bonding between surface metals and acidic organic ligands, particularly with those associated with aromatic structures (Kaiser and Guggenberger 2000). Once DOM is sorbed on mineral surfaces, the desorption of a large part of sorbed DOM is almost fully irreversible under conditions similar to those of adsorption, but it also depends on the surface properties of the sorbate (Kaiser and Guggenberger 2000). A spectroscopic study showed that there is an enormous complexity of the OC functionalities and various inorganic components in the organo-mineral assemblages and interfaces, and it is likely that no single binding mechanism could be accountable for the organic C stored at the micron scale (Solomon et al. 2012). The researchers in this study suggested that the apparent C sequestration at this scale was due to both the cumulative result of physical protection and heterogeneous binding mechanisms (i.e., ion exchange, hydrogen bonding, and hydrophobic bonding) on silicate clay organic complexes and adsorption on external and internal surfaces of clay minerals. Another study reported that sorbed SOM may undergo changes in configuration or may migrate into intra-particle spaces with time after sorption (Kaiser, Mikutta and Guggenberger

2007). These researchers investigated the effects of the residence time of SOM sorbed onto ferrihydrite and goethite and found that with increasing residence time, SOM sorbed to porous minerals becomes decreasingly desorbable because of formation of additional chemical bonds to the surface. Additional studies are needed to investigate different SOM binding mechanisms to minerals and the effects of climate-change variables on these mechanisms.

The soil mineral-sorbed OM interface is very important in terms of understanding SOM stability in natural heterogeneous systems. This interface has been conceptualized as a discrete zonal sequence (Kleber, Sollins and Sutton 2007). In the first contact zone, either stable inner-sphere complexes are formed as a result of ligand exchange between organic functional groups and mineral surface hydroxyls, or proteinaceous materials unfold upon adsorption increasing adhesive strength by adding hydrophobic interactions to electrostatic binding. The second zone is formed when exposed hydrophobic portions of amphiphilic molecules of the first contact zone are shielded from the polar aqueous phase through association with hydrophobic moieties of other amphiphilic molecules. The components of the second zone may exchange more easily with the surrounding soil solution than those in the contact zone, but are still retained with considerable force. The third zone, or kinetic zone, contains organic molecules forming an outer region that is loosely retained by cation bridging, hydrogen bonding, and other interactions. The authors of this model (Kleber, Sollins and Sutton 2007) claimed that the zonal concept of organo-mineral interactions offerred a new basis for understanding and predicting the retention of organic compounds, including contaminants, in soils and sediments. While the zonal concept is extremely useful (apparently, it is the only model that provides a clear picture of the solid surface-solution interface during OM adsorption), additional experimental confirmation of this model is needed. In addition, the implications of this model for stability of SOM during climate change, especially the impact of higher heat levels on the binding processes in the different zones, are not well understood.

Spectroscopic investigations of the SOM spatial heterogeneity at the molecular scale and the nano-scale demonstrate the existence of highly variable, spatially distinct, micro- and nano-C repository zones, where OC is sequestered in agglomerated organo-mineral assemblages (Lehmann et al. 2008; Solomon et al. 2012). These submicron-C repositories have considerably different compositions, indicating a high degree of spatial heterogeneity at the micrometer scale (Solomon et al. 2012).

The spectroscopic investigations also showed that the interfacial chemistry of the organo-mineral assemblages was extremely complex, ranging from Ca, Fe, and Al ions; Fe and Al oxides; hydroxides and oxyhydroxides; to phyllosilicates, which could provide a variety of polyvalent cations, hydroxyl surface functional groups, and edge sites that can bind organic compounds (Solomon et al. 2012). Given the complex nature of the interface between the mineral surface andsorbed SOM, continued investigation of the relationships is definitely an area of research that needs further development (Petridis et al. 2014). In a recent paper, the authors (Petridis et al. 2014) examined the nano-scale structure of a model interface by depositing films of SOM compounds of contrasting chemical character, hydrophilic glucose and amphiphilic stearic acid, onto the surface of an aluminum oxide (a common mineral in soils). They found that glucose molecules reside in a layer between the aluminum oxide and the stearic acid (Petridis et al. 2014). Similar studies involving different minerals, SOM compounds, and climate change relevant conditions are definitely warranted.

The idea of co-precipitation of SOM with Fe oxides was explored in a recent study (Eusterhues et al. 2011). These authors argued that, in soil and water, ferrihydrite frequently forms in the presence of DOM, and this disturbed crystal growth and gives rise to co-precipitation of SOM with ferrihydrite. To compare the fraction of OM co-precipitated with ferrihydite with the fraction involved in adsorption onto pristine ferrihydrite surfaces, the researchers prepared samples of ferrihydrite associated with OM via adsorption and co-precipitation using a forest-floor extract or a sulfonated lignin. They used ^{13}C CPMAS nuclear magnetic resonance, Fourier transform infrared spectroscopy, and analysis of hydrolyzable neutral polysaccharides to study the ferrihydrite-OM associations. They found that, relative to the original forest-floor extract, the ferrihydrite-associated OM is enriched in polysaccharides, especially when adsorption took place. They also found that mannose and glucose are bound preferentially to ferrihydrite, while fucose, arabinose, xylose, and galactose accumulated in the supernatant. This fractionation of sugar monomers is more pronounced during co-precipitation and led to an enhanced ratio of (galactose + mannose)/(arabinose + xylose). Experiments with lignin revealed that the ferrihydrite-associated material is enriched in its aromatic components but had a lower ratio of phenolic C to aromatic C than the original lignin. A compositional difference between the adsorbed and co-precipitated lignin is obvious from a higher contribution of methoxy C in the co-precipitated material. Because co-precipitated SOM will behave differently than adsorbed SOM, a full array of studies are needed to gain information about SOM co-precipitation with soil minerals of different types and under conditions relevant to climate change.

3.3. Temperature Sensitivity

The temperature sensitivity of SOM is a key factor determining the response of the terrestrial C balance to global warming (von Lutzow and Kogel-Knabner 2009). Global climate C-cycle models predict acceleration of SOC losses to the atmosphere with warming, but the size of this feedback is poorly understood (Hopkins et al. 2012). Mainly for this reason and because the response of SOM decomposition to increasing temperature is a critical aspect of ecosystem responses to global change (Conant et al. 2011), many studies have addressed this topic over the last two to three years. However, despite intensive research, a consensus on the effect of temperature on SOM mineralization has not yet emerged (Davidson and Janssens 2006) and remains elusive (Ghee et al. 2013). A better understanding of the relationship between the rate and extent of SOM decomposition and soil temperature is required to make predictions of the impact of climate-change variables on SOM responses at different spatial and temporal scales.

Global climate change may induce accelerated SOM decomposition through increased soil temperature and other important changes, which collectively impact the C balance in soils. Soil C decomposition is sensitive to changes in temperature, and even small increases in temperature may prompt large releases of C from soils (Conant et al. 2008). The following related topics on temperature sensitivity of SOM were covered in other studies conducted over the last 3 years:

- Examination of various soil decomposition and chemical characteristics and their relationship to SOM temperature sensitivity (Haddix et al. 2011).
- Development of a new conceptual model that explicitly identifies the processes controlling soil OM availability for decomposition and allows a description of the factors regulating SOM decomposition under different circumstances (Conant et al. 2011).
- Studies of the vulnerability of soil C that is years-to-decades old, which makes up a large fraction of total soil C in forest soils globally, to warming (Hopkins, Torn and Trumbore 2012; Fissore et al. 2013), which imply that a major portion of soil C may become a source of

atmospheric CO_2 under global warming in the 21st century (Li et al. 2013).
- Comparison of short-term and seasonal responses of soil respiration to a shifting thermal environment and variable substrate availability (Tucker et al. 2013).
- Studies of the relationship between the activation energy of decay of SOM and C and N stoichiometry, and how that can alter the relative availability of C and N as temperature changes (Billings and Ballantyne 2013).
- Another important concept that is insufficiently explored in current investigations of SOM responses to temperature change is the complete range of responses for how warming may change microbial resource demands, physiology, community structure, and total biomass (Billings and Ballantyne 2013).
- Investigations of the temperature sensitivities of basal respiration (partitioned into recent and older SOM sources) and of additional SOM mineralization associated with the addition of labile C to soil (priming effects) (Ghee et al. 2013).

Future research efforts should be focused on the following scientific hypotheses with the overall objective to improve our understanding on temperature sensitivity of SOM mineralization:

1) Several environmental constraints will obscure the intrinsic temperature sensitivity of substrate decomposition, causing lower observed "apparent'" temperature sensitivity, and these constraints may, themselves, be sensitive to climate (Davidson and Janssens 2006). This hypothesis addresses the effects of environmental constraints on what is called "apparent vs. real" SOM temperature sensitivity. Several environmental constraints obscure the intrinsic temperature sensitivity of substrate decomposition, causing lower observed apparent temperature sensitivity (Davidson and Janssens 2006). In addition, factors controlling long-term temperature sensitivity of SOM decomposition are more complex due to the protective effect of the mineral matrix and thus remain as a central question (Wagai et al. 2013).
2) Physical separation or compartmentalization of substrates and decomposers in the soil matrix will decrease SOM sensitivity to temperature (Plante et al. 2009). In this study, the authors reported

that the overall CO_2 efflux increased with temperature, but responses among physical protection treatments were not consistently different. Because the hypothesized attenuation of temperature sensitivity was not detected in this study, these authors concluded that, although compartmentalization of substrates and decomposers is known to reduce the decomposability of SOM labile pool, the sensitivity is probably driven by the thermodynamics of biochemical reactions as expressed by Arrhenius-type equations (Plante et al. 2009). Additional studies are needed to investigate the effects of physical separation of substrates and decomposers on SOM sensitivity to temperature.

3) The temperature response of the processes that control substrate availability, depolymerization, microbial efficiency, and enzyme production will be important in determining the fate of SOM in a warmer world (Conant et al. 2011). For a wide range of forest soils, the supply of labile substrate, controlled through extended incubation and glucose additions, exerted a strong influence on the magnitude of SOC decomposition in response to warming and showed that substrate supply can play a strong role in determining the temperature response of decomposing SOC (Fissore, Giardina and Kolka 2013). Fissore and others (2013) concluded that, because substrate supply was likely to vary both spatially and temporally, these findings have important implications for SOC processing in natural systems.

4) The temperature sensitivity will differ between freshly added organic matter and bulk soil C. The addition of fresh organic matter will stimulate the decomposition of SOM and this priming effect would be temperature dependent (Thiessen et al. 2013). The results presented in this recent study (Thiessen et al. 2013) disagreed with the view of a simple physico-chemically derived substrate-temperature sensitivity relationship of decomposition and the authors concluded that an explicit consideration of microbial processes, such as growth and priming effects, is required to address the issues raised above.

Other unresolved issues, based on the current literature, formulated as hypotheses that should be tested in future studies are:

a. *The labile pool will be more sensitive and responsive to global climate change than the recalcitrant pool.* Contradictory results are presented in the current literature from a variety of studies conducted to answer the question which SOM pool, the labile or the recalcitrant pool, is more sensitive to increasing temperatures, and whether or not the mineralization response to temperature depends on SOM mineralization rate. As soil microflora are considered to be "functionally omnipotent" (i.e, able to decompose any SOM compounds), the temperature dependence of stable SOM pools is the central issue that determines C stocks and stock changes under global warming (von Lutzow and Kogel-Knabner 2009). The impacts of climate warming on decomposition dynamics have not been resolved due to apparently contradictory results from field and lab experiments, most of which have focused on labile C with short turnover time although the majority of total soil C stocks are comprised of OC with turnover times of decades to centuries (Conant et al. 2011). The temperature sensitivity of labile SOM decomposition could either be greater than, less than, or equivalent to that of resistant SOM (Conant et al. 2008). The initial assumption was that soil labile C was more sensitive to temperature variation, whereas SOM resistant components were insensitive and unresponsive to increasing temperature and global warming. However, kinetic theory based on chemical reactions suggests that older, more-resistant C pools may be more temperature sensitive (Conant et al. 2008). Studies have shown that the temperature sensitivity for resistant SOM pools was not significantly different from that of the labile SOM, and some authors believe that both these pools will respond similarly to global warming (Fang et al. 2005). The results of another study suggested that the temperature sensitivity of resistant SOM pools was greater than that for labile SOM and that global change-driven soil C losses may be greater than previously estimated (Conant et al. 2008). Finally, the results of a long-term study conducted with soils from across Europe demonstrated that temperature response was greater in those organic compounds that have a greater mineralization rate (i.e., stable SOM had a higher temperature sensitivity than the labile SOM) (Lefevre et al. 2014). Because of the contradictory results, additional studies are needed to address the issue of whether the mineralization response to temperature depends on the SOC mineralization rate under different soil and climate-change conditions.

2) *Because SOM is a complex mixture of different organic compounds, each of these compounds will exhibit distinct temperature sensitivity making the overall SOM response variable and time-dependent.* Much of the work conducted so far is based on an implicit assumption that

soil C pools are composed of organic matter pools with uniform temperature sensitivities (Conant et al. 2008). However, as it is clearly presented above, SOM is a complex mixture of different organic compounds, each exhibiting, potentially, distinct temperature sensitivity. Unraveling the feedback to climate change is particularly difficult because diverse soil organic compounds exhibit a wide range of kinetic properties, which determine the intrinsic temperature sensitivity of their decomposition (Davidson and Janssens 2006). The question then is about uniform vs. non-uniform temperature sensitivities, and there is a need to focus research on controls over temperature sensitive SOM stabilization and destabilization processes of individual components (i.e., organic compounds) as a basis for understanding kinetic properties of key chemical reactions that determine SOM pool sizes and turnover rates. A recent laboratory incubation study of forest SOM and fresh litter material combined with nuclear magnetic resonance spectroscopy measurements was conducted to make the connection between SOM chemical composition and temperature sensitivity (Erhagen et al. 2013). The results indicated that temperature response of the fresh litter was directly related to the chemical composition of the constituent organic matter, and it decreased with increasing proportions of aromatic and O-aromatic compounds and increased with increased contents of alkyl- and O-alkyl carbons. A more detailed characterization of the ^{13}C aromatic region revealed considerable differences in the aromatic region between litter and SOM, suggesting that the temperature response of decomposition differed between litter and SOM, and that the temperature response of soil decomposition processes can thus be described by the chemical composition of its constituent organic matter (Erhagen et al. 2013).

3) *Other factors (such as soil moisture content, sampling method, incubation time) and their interactions will be influential in controlling temperature response of SOM decomposition.* Temperature and moisture are primary environmental drivers of SOM decomposition, and an improved understanding of how they interact is needed (Gabriel and Kellman 2014). Another study reported that SOM decomposition or soil basal respiration rate is significantly affected by changes in SOM components associated with soil depth, sampling method, and incubation time (Fang et al. 2005).

REFERENCES

Albaladejo J, R Ortiz, N Garcia-Franco, AR Navarro, M Almagro, JG Pintado, and M Martinez-Mena. 2013. "Land Use and Climate Change Impacts on Soil Organic Carbon Stocks in Semi-Arid Spain." Journal of Soils and Sediments 13:265-77.

Altevogt AS and PR Jaffe. 2005. "Modeling the Effects of Gas Phase CO_2 Intrusion on the Biogeochemistry of Variably Saturated Soils." Water Resour. Res. 41.

Ananyeva K, W Wang, AJM Smucker, ML Rivers, and AN Kravchenko. 2013. "Can Intra-Aggregate Pore Structures Affect the Aggregate's Effectiveness in Protecting Carbon?" Soil Biology & Biochemistry 57:868-75.

Angélil O, DA Stone, M Tadross, F Tummon, M Wehner, and R Knutti. 2014. "Attribution of Extreme Weather to Anthropogenic Greenhouse Gas Emissions: Sensitivity to Spatial and Temporal Scales." Geophysical Research Letters:2014GL059234.

Arvidson RS and FT Mackenzie. 1999. "The Dolomite Problem: Control of Precipitation Kinetics by Temperature and Saturation State." American Journal of Science 299:257-88.

Ballantyne AP, CB Alden, JB Miller, PP Tans, and JWC White. 2012. "Increase in Observed Net Carbon Dioxide Uptake by Land and Oceans During the Past 50 Years." Nature 488:70-+.

Baron JS, TM Schmidt, and MD Hartman. 2009. "Climate-Induced Changes in High Elevation Stream Nitrate Dynamics." Global Change Biology 15:1777-89.

Bea SA, SA Wilson, KU Mayer, GM Dipple, IM Power, and P Gamazo. 2012. "Reactive Transport Modeling of Natural Carbon Sequestration in Ultramafic Mine Tailings." Vadose Zone Journal 11.

Bearat H, MJ McKelvy, AVG Chizmeshya, D Gormley, R Nunez, RW Carpenter, K Squires, and GH Wolf. 2006. "Carbon Sequestration Via Aqueous Olivine Mineral Carbonation: Role of Passivating Layer Formation." Environ. Sci. Technol. 40:4802-08.

Beaulieu E, Y Godderis, Y Donnadieu, D Labat, and C Roelandt. 2012. "High Sensitivity of the
Continental-Weathering Carbon Dioxide Sink to Future Climate Change." Nature Climate Change 2:346- 49.

Beinlich A and H Austrheim. 2012. "In Situ Sequestration of Atmospheric CO_2 at Low Temperature and Surface Cracking of Serpentinized Peridotite in Mine Shafts." Chemical Geology 332:32-44.

Bellini G, ME Sumner, DE Radcliffe, and NP Qafoku. 1996. "Anion Transport through Columns of Highly Weathered Acid Soil: Adsorption and Retardation." Soil Sci. Soc. Am. J. 60:132-37.

Billings SA and F Ballantyne. 2013. "How Interactions between Microbial Resource Demands, Soil Organic Matter Stoichiometry, and Substrate Reactivity Determine the Direction and Magnitude of Soil Respiratory Responses to Warming." Global Change Biology 19:90-102.

Birkholzer JT, JA Apps, L Zheng, Y Zhang, T Xu, and C-F Tsang. 2008. "Research Project on CO_2 Geological Storage and Groundwater Resources: Water Quality Effects Caused by CO_2 Intrusion into Shallow Groundwater." Vol 2011, pp. 450. Lawrence Berkeley National Laboratory.

Buettner S, M Kramer, O Chadwick, and A Thompson. 2014. "Mobilization of Colloidal Carbon During Iron Reduction in Basaltic Soils." Geoderma 221-222:139-45.

Burns RG, JL DeForest, J Marxsen, RL Sinsabaugh, ME Stromberger, MD Wallenstein, MN Weintraub, and A Zoppini. 2013. "Soil Enzymes in a Changing Environment: Current Knowledge and Future Directions." Soil Biology & Biochemistry 58:216-34.

Caldwell RJ, S Gangopadhyay, J Bountry, Y Lai, and MM Elsner. 2013. "Statistical Modeling of Daily and Subdaily Stream Temperatures: Application to the Methow River Basin, Washington." Water Resources Research 49:4346-61.

Campbell KM, RK Kukkadapu, NP Qafoku, A Peacock, E Lesher, L Figueroa, KH Williams, MJ Wilkins, CT Resch, JA Davis, and PE Long. 2012. "Characterizing the Extent and Role of Natural Subsurface Bioreduction in a Uranium-Contaminated Aquifer." Appl. Geochem. 27:1499-511.

Cao JH, DX Yuan, C Groves, F Huang, H Yang, and Q Lu. 2012. "Carbon Fluxes and Sinks: The Consumption of Atmospheric and Soil CO_2 by Carbonate Rock Dissolution." Acta Geologica SinicaEnglish Edition 86:963-72.

Chabbi A, I Kogel-Knabner, and C Rumpel. 2009. "Stabilised Carbon in Subsoil Horizons Is Located in Spatially Distinct Parts of the Soil Profile." Soil Biology & Biochemistry 41:256-61.

Conant RT, MG Ryan, GI Agren, HE Birge, EA Davidson, PE Eliasson, SE Evans, SD Frey, CP Giardina, FM Hopkins, R Hyvonen, MUF

Kirschbaum, JM Lavallee, J Leifeld, WJ Parton, JM Steinweg, MD Wallenstein, JAM Wetterstedt, and MA Bradford. 2011. "Temperature and Soil Organic Matter Decomposition Rates - Synthesis of Current Knowledge and a Way Forward." Global Change Biology 17:3392-404.

Conant RT, JM Steinweg, ML Haddix, EA Paul, AF Plante, and J Six. 2008. "Experimental Warming Shows That Decomposition Temperature Sensitivity Increases with Soil Organic Matter Recalcitrance." Ecology 89:2384-91.

Crosbie RS, BR Scanlon, FS Mpelasoka, RC Reedy, JB Gates, and L Zhang. 2013. "Potential Climate Change Effects on Groundwater Recharge in the High Plains Aquifer, USA." Water Resources Research 49:3936-51.

Davidson EA and IA Janssens. 2006. "Temperature Sensitivity of Soil Carbon Decomposition and Feedbacks to Climate Change." Nature 440:165-73.

Davidson EA, S Samanta, SS Caramori, and K Savage. 2012. "The Dual Arrhenius and MichaelisMenten Kinetics Model for Decomposition of Soil Organic Matter at Hourly to Seasonal Time Scales." Global Change Biology 18:371-84.

Dere AL, TS White, RH April, B Reynolds, TE Miller, EP Knapp, LD McKay, and SL Brantley. 2013. "Climate Dependence of Feldspar Weathering in Shale Soils Along a Latitudinal Gradient." Geochimica et Cosmochimica Acta 122:101-26.

Dixon JL and F von Blanckenburg. 2012. "Soils as Pacemakers and Limiters of Global Silicate Weathering." Comptes Rendus Geoscience 344:597-609.

Dong H. 2010. "Mineral-Microbe Interactions: A Review." Front. Earth Sci. China 4:127-47.

Douglas S and TJ Beveridge. 1998. "Mineral Formation by Bacteria in Natural Microbial Communities." FEMS Microbiol. Ecol. 26:79-88.

Erhagen B, M Oquist, T Sparrman, M Haei, U Ilstedt, M Hedenstrom, J Schleucher, and MB Nilsson. 2013. "Temperature Response of Litter and Soil Organic Matter Decomposition Is Determined by Chemical Composition of Organic Material." Global Change Biology 19:3858-71.

Eusterhues K, T Rennert, H Knicker, I Kogel-Knabner, KU Totsche, and U Schwertmann. 2011. "Fractionation of Organic Matter Due to Reaction with Ferrihydrite: Coprecipitation Versus Adsorption." Environmental Science & Technology 45:527-33.

Evans RD, A Koyama, DL Sonderegger, TN Charlet, BA Newingham, LF Fenstermaker, B Harlow, VL Jin, K Ogle, SD Smith, and RS Nowak. 2014. "Greater Ecosystem Carbon in the Mojave Desert after Ten Years

Exposure to Elevated CO_2." Nature Clim. Change advance online publication.

Falkowski P, RJ Scholes, E Boyle, J Canadell, D Canfield, J Elser, N Gruber, K Hibbard, P Högberg, S Linder, FT Mackenzie, B Moore III, T Pedersen, Y Rosenthal, S Seitzinger, V Smetacek, and W Steffen. 2000. "The Global Carbon Cycle: A Test of Our Knowledge of Earth as a System." Science 290:291-96.

Fang CM, P Smith, JB Moncrieff, and JU Smith. 2005. "Similar Response of Labile and Resistant Soil Organic Matter Pools to Changes in Temperature." Nature 433:57-59.

Feng WT, AF Plante, AK Aufdenkampe, and J Six. 2014. "Soil Organic Matter Stability in OrganoMineral Complexes as a Function of Increasing C Loading." Soil Biology & Biochemistry 69:398-405.

Fiantis D, M Nelson, E Van Ranst, J Shamsuddin, and NP Qafoku. 2009. "Chemical Weathering of New Pyroclastic Deposits from Mt. Merapi (Java), Indonesia." Journal of Mountain Science 6:240-54.

Fissore C, CP Giardina, and RK Kolka. 2013. "Reduced Substrate Supply Limits the Temperature Response of Soil Organic Carbon Decomposition." Soil Biology & Biochemistry 67:306-11.

Follett RF, CE Stewart, EG Pruessner, and JM Kimble. 2012. "Effects of Climate Change on Soil Carbon and Nitrogen Storage in the Us Great Plains." Journal of Soil and Water Conservation 67:331-42.

Fontaine S, S Barot, P Barre, N Bdioui, B Mary, and C Rumpel. 2007. "Stability of Organic Carbon in Deep Soil Layers Controlled by Fresh Carbon Supply." Nature 450:277-U10.

Fornara DA, S Steinbeiss, NP McNamara, G Gleixner, S Oakley, PR Poulton, AJ Macdonald, and RD Bardgett. 2011. "Increases in Soil Organic Carbon Sequestration Can Reduce the Global Warming Potential of Long-Term Liming to Permanent Grassland (Vol 17, Pg 1925, 2011)." Global Change Biology 17:2762-62.

Friedlingstein P, RA Houghton, G Marland, J Hackler, TA Boden, TJ Conway, JG Canadell, MR Raupach, P Ciais, and C Le Quere. 2010. "Update on CO_2 Emissions." Nature Geoscience 3:811-12. Gabriel, CE, and L Kellman. 2014. "Investigating the Role of Moisture as an Environmental Constraint in the Decomposition of Shallow and Deep Mineral Soil Organic Matter of a Temperate Coniferous Soil." Soil Biology & Biochemistry 68:373-84.

Gaillardet J, B Dupre, P Louvat, and CJ Allegre. 1999. "Global Silicate Weathering and CO_2 Consumption Rates Deduced from the Chemistry of Large Rivers." Chemical Geology 159:3-30.

Gangloff S, P Stille, M-C Pierret, T Weber, and F Chabaux. 2014. "Characterization and Evolution of Dissolved Organic Matter in Acidic Forest Soil and Its Impact on the Mobility of Major and Trace Elements (Case of the Strengbach Watershed)." Geochimica et Cosmochimica Acta 130:21-41.

Ghee C, R Neilson, PD Hallett, D Robinson, and E Paterson. 2013. "Priming of Soil Organic Matter Mineralisation Is Intrinsically Insensitive to Temperature." Soil Biology & Biochemistry 66:20-28.

Gislason SR, EH Oelkers, ES Eiriksdottir, MI Kardjilov, G Gisladottir, B Sigfusson, A Snorrason, S Elefsen, J Hardardottir, P Torssander, and N Oskarsson. 2009. "Direct Evidence of the Feedback between Climate and Weathering." Earth and Planetary Science Letters 277:213-22.

Gislason SR, D Wolff-Boenisch, A Stefansson, EH Oelkers, E Gunnlaugsson, H Sigurdardottir, B Sigfusson, WS Broecker, JM Matter, M Stute, G Axelsson, and T Fridriksson. 2010. "Mineral Sequestration of Carbon Dioxide in Basalt: A Pre-Injection Overview of the Carbfix Project." Int. J. Greenhouse Gas Control 4:537-45.

Gruneberg E, I Schoning, D Hessenmoller, ED Schulze, and WW Weisser. 2013. "Organic Layer and Clay Content Control Soil Organic Carbon Stocks in Density Fractions of Differently Managed German Beech Forests." Forest Ecology and Management 303:1-10.

Guimaraes DV, MIS Gonzaga, TO da Silva, TL da Silva, ND Dias, and MIS Matias. 2013. "Soil Organic Matter Pools and Carbon Fractions in Soil under Different Land Uses." Soil & Tillage Research 126:177-82.

Haddix ML, AF Plante, RT Conant, J Six, JM Steinweg, K Magrini-Bair, RA Drijber, SJ Morris, and EA Paul. 2011. "The Role of Soil Characteristics on Temperature Sensitivity of Soil Organic Matter." Soil Science Society of America Journal 75:56-68.

Hall SJ and WL Silver. 2013. "Iron Oxidation Stimulates Organic Matter Decomposition in Humid Tropical Forest Soils." Global Change Biology 19:2804-13.

Harrison AL, IM Power, and GM Dipple. 2013. "Accelerated Carbonation of Brucite in Mine Tailings for Carbon Sequestration." Environmental Science & Technology 47:126-34.

Hartmann J, AJ West, P Renforth, P Kohler, CL De La Rocha, DA Wolf-Gladrow, HH Durr, and J Scheffran. 2013. "Enhanced Chemical

Weathering as a Geoengineering Strategy to Reduce Atmospheric Carbon Dioxide, Supply Nutrients, and Mitigate Ocean Acidification." Reviews of Geophysics 51:113- 49.

Harvey OR, NP Qafoku, KJ Cantrell, G Lee, JE Amonette, and CF Brown. 2013. "Geochemical Implications of Gas Leakage Associated with Geologic CO_2 Storage-a Qualitative Review." Environmental Science & Technology 47:23-36.

Hernes PJ, K Kaiser, RY Dyda, and C Cerli. 2013. "Molecular Trickery in Soil Organic Matter: Hidden Lignin." Environmental Science & Technology 47:9077-85.

Hodgkins SB, MM Tfaily, CK McCalley, TA Logan, PM Crill, SR Saleska, VI Rich, and JP Chanton. 2014. "Changes in Peat Chemistry Associated with Permafrost Thaw Increase Greenhouse Gas Production." Proceedings of the National Academy of Sciences.

Hopkins FM, MS Torn, and SE Trumbore. 2012. "Warming Accelerates Decomposition of Decades-Old Carbon in Forest Soils." Proceedings of the National Academy of Sciences of the United States of America 109:E1753-E61.

Houghton RA, F Hall, and SJ Goetz. 2009. "Importance of Biomass in the Global Carbon Cycle." Journal of Geophysical Research-Biogeosciences 114:13.

IPCC Climate Change. 2007. The Physical Science Basis Cambridge University Press.

Isaak, DJ, CC Muhlfeld, AS Todd, R Al-Chokhachy, J Roberts, JL Kershner, KD Fausch, and SW Hostetler. 2012. "The Past as Prelude to the Future for Understanding 21st-Century Climate Effects on Rocky Mountain Trout." Fisheries 37:542-56.

Isaak DJ, S Wollrab, D Horan, and G Chandler. 2012. "Climate Change Effects on Stream and River Temperatures across the Northwest Us from 1980-2009 and Implications for Salmonid Fishes." Climatic Change 113:499-524.

Ishihara S, P Sahoo, K Deguchi, S Ohki, M Tansho, T Shimizu, J Labuta, JP Hill, K Ariga, K Watanabe, Y Yamauchi, S Suehara, and N Iyi. 2013. "Dynamic Breathing of CO_2 by Hydrotalcite." Journal of the American Chemical Society.

Jimenez-Lopez C and CS Romanek. 2004. "Precipitation Kinetics and Carbon Isotope Partitioning of Inorganic Siderite at 25 Degrees C and 1 Atm." Geochimica Et Cosmochimica Acta 68:557-71.

Jobbagy EG and RB Jackson. 2000. "The Vertical Distribution of Soil Organic Carbon and Its Relation to Climate and Vegetation." Ecological Applications 10:423-36.

Kaiser K and G Guggenberger. 2000. "The Role of Dom Sorption to Mineral Surfaces in the Preservation of Organic Matter in Soils." Organic Geochemistry 31:711-25.

Kaiser K and G Guggenberger. 2003. "Mineral Surfaces and Soil Organic Matter." European Journal of Soil Science 54:219-36.

Kaiser K and G Guggenberger. 2007. "Sorptive Stabilization of Organic Matter by Microporous

Goethite: Sorption into Small Pores Vs. Surface Complexation." European Journal of Soil Science 58:45- 59.

Kaiser K and K Kalbitz. 2012. "Cycling Downwards - Dissolved Organic Matter in Soils." Soil Biology & Biochemistry 52:29-32.

Kaiser K, R Mikutta, and G Guggenberger. 2007. "Increased Stability of Organic Matter Sorbed to Ferrihydrite and Goethite on Aging." Soil Science Society of America Journal 71:711-19.

Kakizawa M, A Yamasaki, and Y Yanagisawa. 2001. "A New CO_2 Disposal Process Via Artificial Weathering of Calcium Silicate Accelerated by Acetic Acid." Energy 26:341-54.

Kalbitz K, D Schwesig, J Rethemeyer, and E Matzner. 2005. "Stabilization of Dissolved Organic Matter by Sorption to the Mineral Soil." Soil Biology & Biochemistry 37:1319-31.

Kelemen PB, and J Matter. 2008. "In Situ Carbonation of Peridotite for CO_2 Storage." Proceedings of the National Academy of Sciences of the United States of America 105:17295-300.

Kelemen PB, J Matter, EE Streit, JF Rudge, WB Curry, and J Blusztajn. 2011. "Rates and Mechanisms of Mineral Carbonation in Peridotite: Natural Processes and Recipes for Enhanced, in Situ CO_2 Capture and Storage." in Annual Review of Earth and Planetary Sciences, Vol 39, eds. R Jeanloz and KH Freeman, Vol 39, pp. 545-76. Annual Reviews, Palo Alto.

Kelleher BP and AJ Simpson. 2006. "Humic Substances in Soils: Are They Really Chemically Distinct?" Environmental Science & Technology 40:4605-11.

Kenward PA, RH Goldstein, LA GonzÁLez, and JA Roberts. 2009. "Precipitation of Low-Temperature Dolomite from an Anaerobic Microbial Consortium: The Role of Methanogenic Archaea." Geobiology 7:556-65.

Kharaka YK, JJ Thordsen, E Kakouros, G Ambats, WN Herkelrath, SR Beers, JT Birkholzer, JA Apps, NF Spycher, LE Zheng, RC Trautz, HW Rauch, and KS Gullickson. 2010. "Changes in the Chemistry of Shallow Groundwater Related to the 2008 Injection of CO_2 at the Zert Field Site, Bozeman, Montana." Environ. Earth Sci. 60:273-84.

King HE, O Plumper, and A Putnis. 2010. "Effect of Secondary Phase Formation on the Carbonation of Olivine." Environmental Science & Technology 44:6503-09.

Kinyangi J, D Solomon, BI Liang, M Lerotic, S Wirick, and J Lehmann. 2006. "Nanoscale Biogeocomplexity of the Organomineral Assemblage in Soil: Application of Stxm Microscopy and C 1sNexafs Spectroscopy." Soil Science Society of America Journal 70:1708-18.

Kleber M. 2010. "What Is Recalcitrant Soil Organic Matter?" Environmental Chemistry 7:320-32.

Kleber, M, and MG Johnson. 2010. "Advances in Understanding the Molecular Structure of Soil Organic Matter: Implications for Interactions in the Environment." in Advances in Agronomy, Vol 106, ed. DL Sparks, Vol 106, pp. 77-142. Elsevier Academic Press Inc, San Diego.

Kleber M, PS Nico, AF Plante, T Filley, M Kramer, C Swanston, and P Sollins. 2011. "Old and Stable Soil Organic Matter Is Not Necessarily Chemically Recalcitrant: Implications for Modeling Concepts and Temperature Sensitivity." Global Change Biology 17:1097-107.

Kleber M, P Sollins, and R Sutton. 2007. "A Conceptual Model of Organo-Mineral Interactions in Soils: Self-Assembly of Organic Molecular Fragments into Zonal Structures on Mineral Surfaces." Biogeochemistry 85:9-24.

Köhler P, J Hartmann, and DA Wolf-Gladrow. 2010. "Geoengineering Potential of Artificially Enhanced Silicate Weathering of Olivine." Proceedings of the National Academy of Sciences 107:20228- 33.

Köhler P, J Hartmann, and DA Wolf-Gladrow. 2011. "Reply to Schuiling Et Al.: Different Processes at Work." Proceedings of the National Academy of Sciences 108:E42.

Kojima T, A Nagamine, N Ueno, and S Uemiya. 1997. "Absorption and Fixation of Carbon Dioxide by Rock Weathering." Energy Conversion and Management 38, Supplement:S461-S66.

Kramer MG, J Sanderman, OA Chadwick, J Chorover, and PM Vitousek. 2012. "Long-Term Carbon Storage through Retention of Dissolved

Aromatic Acids by Reactive Particles in Soil." Global Change Biology 18:2594-605.

Krevor SC, and KS Lackner. 2009. "Enhancing Process Kinetics for Mineral Carbon Sequestration." Energy Procedia 1:4867-71.

Lackner KS, DP Butt, and CH Wendt. 1997. "Progress on Binding CO_2 in Mineral Substrates." Energy Conversion and Management 38, Supplement:S259-S64.

Lackner KS, CH Wendt, DP Butt, EL Joyce Jr, and DH Sharp. 1995. "Carbon Dioxide Disposal in Carbonate Minerals." Energy 20:1153-70.

Lal R. 2004. "Soil Carbon Sequestration Impacts on Global Climate Change and Food Security." Science 304:1623.

Lal R. 2008. "Sequestration of Atmospheric CO_2 in Global Carbon Pools." Energy & Environmental Science 1:86-100.

Lal R. 2013. "Soil Carbon Management and Climate Change." Carbon Management 4:439-62.

Lalonde K, A Mucci, A Ouellet, and Y Gelinas. 2012. "Preservation of Organic Matter in Sediments Promoted by Iron." Nature 483:198-200.

Langmuir D. 1997. Aqueous Environmental Geochemistry. Prentice-Hall Inc., New Jersey. Laudon, H, PJ Dillon, MC Eimers, RG Semkin, and DS Jeffries. 2004. "Climate-Induced Episodic Acidification of Streams in Central Ontario." Environmental Science & Technology 38:6009-15.

Lefevre R, P Barre, FE Moyano, BT Christensen, G Bardoux, T Eglin, C Girardin, S Houot, T Katterer, F van Oort, and C Chenu. 2014. "Higher Temperature Sensitivity for Stable Than for Labile Soil Organic Carbon - Evidence from Incubations of Long- Term Bare Fallow Soils." Global Change Biology 20:633- 40.

Lehmann J, D Solomon, J Kinyangi, L Dathe, S Wirick, and C Jacobsen. 2008. "Spatial Complexity of Soil Organic Matter Forms at Nanometre Scales." Nature Geoscience 1:238-42.

Li DJ, C Schaedel, ML Haddix, EA Paul, R Conant, JW Li, JZ Zhou, and YQ Luo. 2013. "Differential Responses of Soil Organic Carbon Fractions to Warming: Results from an Analysis with Data Assimilation." Soil Biology & Biochemistry 67:24-30.

Li S, XX Lu, and RT Bush. 2014. "Chemical Weathering and CO_2 Consumption in the Lower Mekong River." Science of The Total Environment 472:162-77.

Lin YP and PC Singer. 2005. "Effects of Seed Material and Solution Composition on Calcite Precipitation." Geochimica Et Cosmochimica Acta 69:4495-504.

Lin YP, PC Singer, and GR Aiken. 2005. "Inhibition of Calcite Precipitation by Natural Organic Material: Kinetics, Mechanism, and Thermodynamics." Environmental Science & Technology 39:6420- 28.

Little MG and RB Jackson. 2010. "Potential Impacts of Leakage from Deep CO_2 Geosequestration on Overlying Freshwater Aquifers." Environ. Sci. Technol. 44:9225-32.

Lopez-Sangil L and P Rovira. 2013. "Sequential Chemical Extractions of the Mineral-Associated Soil Organic Matter: An Integrated Approach for the Fractionation of Organo-Mineral Complexes." Soil Biology & Biochemistry 62:57-67.

Loring JS, CJ Thompson, Z Wang, AG Joly, DS Sklarew, HT Schaef, ES Ilton, KM Rosso, and AR Felmy. 2011. "In Situ Infrared Spectroscopic Study of Forsterite Carbonation in Wet Supercritical CO_2." Environ. Sci. Technol. 45:6204-10.

Loring JS, CJ Thompson, CY Zhang, ZM Wang, HT Schaef, and KM Rosso. 2012. "In Situ Infrared Spectroscopic Study of Brucite Carbonation in Dry to Water-Saturated Supercritical Carbon Dioxide." Journal of Physical Chemistry A 116:4768-77.

Lu JM, JW Partin, SD Hovorka, and C Wong. 2010. "Potential Risks to Freshwater Resources as a Result of Leakage from CO_2 Geological Storage: A Batch-Reaction Experiment." Environ. Earth Sci. 60:335-48.

Maia C, EH Novotny, TF Rittl, and MHB Hayes. 2013. "Soil Organic Matter: Chemical and Physical Characteristics and Analytical Methods. A Review." Current Organic Chemistry 17:2985-90.

Maity R, M Ramadas, and RS Govindaraju. 2013. "Identification of Hydrologic Drought Triggers from Hydroclimatic Predictor Variables." Water Resources Research 49:4476-92.

Marschner B, S Brodowski, A Dreves, G Gleixner, A Gude, PM Grootes, U Hamer, A Heim, G Jandl, R Ji, K Kaiser, K Kalbitz, C Kramer, P Leinweber, J Rethemeyer, A Schaeffer, MWI Schmidt, L Schwark, and GLB Wiesenberg. 2008. "How Relevant Is Recalcitrance for the Stabilization of Organic Matter in Soils?" Journal of Plant Nutrition and Soil Science-Zeitschrift Fur Pflanzenernahrung Und Bodenkunde 171:91- 110.

Martin MP, TG Orton, E Lacarce, J Meersmans, NPA Saby, JB Paroissien, C Jolivet, L Boulonne, and D Arrouays. 2014. "Evaluation of Modelling Approaches for Predicting the Spatial Distribution of Soil Organic Carbon Stocks at the National Scale." Geoderma 223-225:97-107.

Mayer LM, LL Schick, KR Hardy, R Wagai, and J McCarthy. 2004. "Organic Matter in Small Mesopores in Sediments and Soils." Geochimica Et Cosmochimica Acta 68:3863-72.

McGrail BP, HT Schaef, VA Glezakou, LX Dang, and AT Owen. 2009. "Water Reactivity in the Liquid and Supercritical CO_2 Phase: Has Half the Story Been Neglected?" in Greenhouse Gas Control Technologies 9, eds. J Gale, H Herzog and J Braitsch, Vol 1, pp. 3415-19. Elsevier Science Bv, Amsterdam.

McGrail BP, HT Schaef, AM Ho, YJ Chien, JJ Dooley, and CL Davidson. 2006. "Potential for Carbon Dioxide Sequestration in Flood Basalts." J. Geophys. Res.-Solid Earth 111.

Mills SJ, SA Wilson, GM Dipple, and M Raudsepp. 2010. "The Decomposition of Konyaite: Importance in CO_2 Fixation in Mine Tailings." Mineralogical Magazine 74:903-17.

Milne AE, J Lehmann, D Solomon, and RM Lark. 2011. "Wavelet Analysis of Soil Variation at Nanometre- to Micrometre-Scales: An Example of Organic Carbon Content in a Micro-Aggregate." European Journal of Soil Science 62:617-28.

Moosdorf N, J Hartmann, R Lauerwald, B Hagedorn, and S Kempe. 2011. "Atmospheric CO_2
Consumption by Chemical Weathering in North America." Geochimica Et Cosmochimica Acta 75:7829- 54.

Moosdorf N, P Renforth, and J Hartmann. 2014. "Carbon Dioxide Efficiency of Terrestrial Enhanced Weathering." Environmental Science & Technology.

Mozley PS and SJ Burns. 1993. "Oxygen and Carbon Isotopic Composition of Marine Carbonate Concretions - an Overview - Reply." Journal of Sedimentary Petrology 63:1008-08.

Oldenburg CM and AJA Unger. 2003. "On Leakage and Seepage from Geologic Carbon Sequestration Sites: Unsaturated Zone Attenuation." Vadose Zone J. 2:287-96.

Olsson J, N Bovet, E Makovicky, K Bechgaard, Z Balogh, and SLS Stipp. 2012. "Olivine Reactivity with CO_2 and H2o on a Microscale: Implications for Carbon Sequestration." Geochimica Et Cosmochimica Acta 77:86-97.

Pangle LA, JW Gregg, and JJ McDonnell. 2014. "Rainfall Seasonality and an Ecohydrological Feedback Offset the Potential Impact of Climate Warming on Evapotranspiration and Groundwater Recharge." Water Resources Research 50:1308-21.

Paterson E and A Sim. 2013. "Soil-Specific Response Functions of Organic Matter Mineralization to the Availability of Labile Carbon." Global Change Biology 19:1562-71.

Patterson LA, B Lutz, and MW Doyle. 2013. "Climate and Direct Human Contributions to Changes in Mean Annual Streamflow in the South Atlantic, USA." Water Resources Research 49:7278-91.

Pereira R, C Isabella Bovolo, RGM Spencer, PJ Hernes, E Tipping, A Vieth-Hillebrand, N Pedentchouk, NA Chappell, G Parkin, and T Wagner. 2014. "Mobilization of Optically Invisible Dissolved Organic Matter in Response to Rainstorm Events in a Tropical Forest Headwater River." Geophysical Research Letters 41:2013GL058658.

Petridis L, H Arnbaye, S Jagadamma, SM Kilbey, BS Lokitz, V Lauter, and MA Mayes. 2014. "Spatial Arrangement of Organic Compounds on a Model Mineral Surface: Implications for Soil Organic Matter Stabilization." Environmental Science & Technology 48:79-84.

Plante AF, RT Conant, CE Stewart, K Paustian, and J Six. 2006. "Impact of Soil Texture on the Distribution of Soil Organic Matter in Physical and Chemical Fractions." Soil Science Society of America Journal 70:287-96.

Plante AF, J Six, EA Paul, and RT Conant. 2009. "Does Physical Protection of Soil Organic Matter Attenuate Temperature Sensitivity?" Soil Science Society of America Journal 73:1168-72.

Power IM, SA Wilson, and GM Dipple. 2013. "Serpentinite Carbonation for CO_2 Sequestration." Elements 9:115-21.

Power IM, SA Wilson, DP Small, GM Dipple, W Wan, and G Southam. 2011. "Microbially Mediated Mineral Carbonation: Roles of Phototrophy and Heterotrophy." Environ. Sci. Technol. 45:9061-68.

Prentice IC. 2001. "The Carbon Cycle and Atmospheric Carbon Dioxide". Climate Change 2001: The Scientific Basis: Contribution of Working Group I to the Third Assessment Report of the Intergovernmental Panel on Climate Change / Houghton, J.T. [Edit.] Retrieved 31 May 2012.".

Qafoku NP. 2010. "Terrestrial Nanoparticles and Their Controls on Soil/Geo Processes and Reactions." Adv. Agron. 107 33-91.

Qafoku NP, BN Gartman, RK Kukkadapu, BW Arey, KH Williams, PJ Mouser, SM Heald, JR Bargar, N Janot, S Yabusaki, and PE Long. 2014. "Geochemical and Mineralogical Investigation of Uranium in Multi-Element Contaminated, Organic-Rich Subsurface Sediment." Appl. Geochem. 42:77-85.

Qafoku NP, R Kukkadapu, JP McKinley, BW Arey, SD Kelly, CT Resch, and PE Long. 2009. "Uranium in Framboidal Pyrite from a Naturally Bioreduced Alluvial Sediment " Environ. Sci. Technol. 43:8528-34.

Qafoku NP, ME Sumner, and LT West. 2000. "Mineralogy and Chemistry of Some Variable Charge Subsoils." Communications in Soil Science and Plant Analysis 31:1051-70.

Qafoku NP, E Van Ranst, A Noble, and G Baert. 2004. "Variable Charge Soils: Their Mineralogy, Chemistry and Management." Adv. Agron. 84:159-215.

Qin ZC, Y Huang, and QL Zhuang. 2013. "Soil Organic Carbon Sequestration Potential of Cropland in China." Global Biogeochemical Cycles 27:711-22.

Regnier P, P Friedlingstein, P Ciais, FT Mackenzie, N Gruber, IA Janssens, GG Laruelle, R Lauerwald, S Luyssaert, AJ Andersson, S Arndt, C Arnosti, AV Borges, AW Dale, A Gallego-Sala, Y Godderis, N Goossens, J Hartmann, C Heinze, T Ilyina, F Joos, DE LaRowe, J Leifeld, FJR Meysman, G Munhoven, PA Raymond, R Spahni, P Suntharalingam, and M Thullner. 2013. "Anthropogenic Perturbation of the Carbon Fluxes from Land to Ocean." Nature Geosci 6:597-607.

Reichstein M, M Bahn, P Ciais, D Frank, MD Mahecha, SI Seneviratne, J Zscheischler, C Beer, N Buchmann, DC Frank, D Papale, A Rammig, P Smith, K Thonicke, M van der Velde, S Vicca, A Walz, and M Wattenbach. 2013. "Climate Extremes and the Carbon Cycle." Nature 500:287-95.

Remusat L, PJ Hatton, PS Nico, B Zeller, M Kleber, and D Derrien. 2012. "Nanosims Study of Organic Matter Associated with Soil Aggregates: Advantages, Limitations, and Combination with Stxm." Environmental Science & Technology 46:3943-49.

Renforth P. 2012. "The Potential of Enhanced Weathering in the Uk." International Journal of Greenhouse Gas Control 10:229-43.

Rice CW. 2002. "Carbon in Soil: Why and How? Geotimes, . American Geological Institute."

Ritson JP, NJD Graham, MR Templeton, JM Clark, R Gough, and C Freeman. 2014. "The Impact of Climate Change on the Treatability of Dissolved Organic Matter (Dom) in Upland Water Supplies: A Uk Perspective." Science of The Total Environment 473–474:714-30.

Roberts JA, PC Bennett, LA Gonzalez, GL Macpherson, and KL Milliken. 2004. "Microbial Precipitation of Dolomite in Methanogenic Groundwater." Geology 32:277-80.

Ruiz-Agudo E, K Kudłacz, CV Putnis, A Putnis, and C Rodriguez-Navarro. 2013. "Dissolution and Carbonation of Portlandite [Ca(Oh)2] Single Crystals." Environmental Science & Technology 47:11342- 49.

Ryskov YG, VA Demkin, SA Oleynik, and EA Ryskova. 2008. "Dynamics of Pedogenic Carbonate for the Last 5000 Years and Its Role as a Buffer Reservoir for Atmospheric Carbon Dioxide in Soils of Russia." Global and Planetary Change 61:63-69.

Saldi GD, G Jordan, J Schott, and EH Oelkers. 2009. "Magnesite Growth Rates as a Function of Temperature and Saturation State." Geochimica Et Cosmochimica Acta 73:5646-57.

Salek SS, R Kleerebezem, HM Jonkers, GJ Witkamp, and MCM van Loosdrecht. 2013. "Mineral CO_2 Sequestration by Environmental Biotechnological Processes." Trends in Biotechnology 31:139-46.

Schmidt MWI, MS Torn, S Abiven, T Dittmar, G Guggenberger, IA Janssens, M Kleber, I KogelKnabner, J Lehmann, DAC Manning, P Nannipieri, DP Rasse, S Weiner, and SE Trumbore. 2011. "Persistence of Soil Organic Matter as an Ecosystem Property." Nature 478:49-56.

Schrumpf M, K Kaiser, G Guggenberger, T Persson, I Kogel-Knabner, and ED Schulze. 2013. "Storage and Stability of Organic Carbon in Soils as Related to Depth, Occlusion within Aggregates, and Attachment to Minerals." Biogeosciences 10:1675-91.

Schuiling RD. 2012. "Capturing CO_2 from Air." Proceedings of the National Academy of Sciences of the United States of America 109:E1210-E10.

Schuiling RD and P Krijgsman. 2006. "Enhanced Weathering: An Effective and Cheap Tool to Sequester CO_2." Climatic Change 74:349-54.

Schuiling RD, SA Wilson, and IM Power. 2011. "Enhanced Silicate Weathering Is Not Limited by Silicic Acid Saturation." Proceedings of the National Academy of Sciences of the United States of America 108:E41-E41.

Simpson MJ and AJ Simpson. 2012. "The Chemical Ecology of Soil Organic Matter Molecular Constituents." Journal of Chemical Ecology 38:768-84.

Sissmann O, F Brunet, I Martinez, F Guyot, A Verlaguet, Y Pinquier, and D Daval. 2014. "Enhanced Olivine Carbonation within a Basalt as Compared to Single-Phase Experiments: Reevaluating the Potential of CO_2 Mineral Sequestration." Environmental Science & Technology.

Six J, RT Conant, EA Paul, and K Paustian. 2002. "Stabilization Mechanisms of Soil Organic Matter: Implications for C-Saturation of Soils." Plant and Soil 241:155-76.

Six J, SM Ogle, FJ Breidt, RT Conant, AR Mosier, and K Paustian. 2004. "The Potential to Mitigate Global Warming with No-Tillage Management Is Only Realized When Practised in the Long Term." Global Change Biology 10:155-60.

Solomon D, J Lehmann, J Harden, J Wang, J Kinyangi, K Heymann, C Karunakaran, YS Lu, S Wirick, and C Jacobsen. 2012. "Micro- and Nano-Environments of Carbon Sequestration: Multi-Element StxmNexafs Spectromicroscopy Assessment of Microbial Carbon and Mineral Associations." Chemical Geology 329:53-73.

Solomon D, J Lehmann, J Thies, T Schafer, BQ Liang, J Kinyangi, E Neves, J Petersen, F Luizao, and J Skjemstad. 2007. "Molecular Signature and Sources of Biochemical Recalcitrance of Organic C in Amazonian Dark Earths." Geochimica Et Cosmochimica Acta 71:2285-98.

Song ZL, HL Wang, PJ Strong, ZM Li, and PK Jiang. 2012. "Plant Impact on the Coupled Terrestrial Biogeochemical Cycles of Silicon and Carbon: Implications for Biogeochemical Carbon Sequestration." Earth-Science Reviews 115:319-31.

Stewart CE, K Paustian, RT Conant, AF Plante, and J Six. 2007. "Soil Carbon Saturation: Concept, Evidence and Evaluation." Biogeochemistry 86:19-31.

Stewart CE, K Paustian, RT Conant, AF Plante, and J Six. 2008. "Soil Carbon Saturation: Evaluation and Corroboration by Long-Term Incubations." Soil Biology & Biochemistry 40:1741-50.

Stewart CE, AF Plante, K Paustian, RT Conant, and J Six. 2008. "Soil Carbon Saturation: Linking Concept and Measurable Carbon Pools." Soil Science Society of America Journal 72:379-92.

Stockmann U, MA Adams, JW Crawford, DJ Field, N Henakaarchchi, M Jenkins, B Minasny, AB McBratney, VD de Courcelles, K Singh, I Wheeler, L Abbott, DA Angers, J Baldock, M Bird, PC Brookes, C Chenu, JD Jastrowh, R Lal, J Lehmann, AG O'Donnell, WJ Parton, D Whitehead, and M Zimmermann. 2013. "The Knowns, Known Unknowns and Unknowns of Sequestration of Soil Organic Carbon." Agriculture Ecosystems & Environment 164:80-99.

Strauss J, L Schirrmeister, G Grosse, S Wetterich, M Ulrich, U Herzschuh, and H-W Hubberten. 2013. "The Deep Permafrost Carbon Pool of the Yedoma Region in Siberia and Alaska." Geophysical Research Letters 40:2013GL058088.

Street-Perrott FA and PA Barker. 2008. "Biogenic Silica: A Neglected Component of the Coupled Global Continental Biogeochemical Cycles of Carbon and Silicon." Earth Surface Processes and Landforms 33:1436-57.

Sutton R and G Sposito. 2005. "Molecular Structure in Soil Humic Substances: The New View." Environmental Science & Technology 39:9009-15.

Tagliabue A, O Aumont, and L Bopp. 2014. "The Impact of Different External Sources of Iron on the Global Carbon Cycle." Geophysical Research Letters 41:2013GL059059.

ten Berge HFM, HG van der Meer, JW Steenhuizen, PW Goedhart, P Knops, and J Verhagen. 2012. "Olivine Weathering in Soil, and Its Effects on Growth and Nutrient Uptake in Ryegrass (Lolium Perenne L.): A Pot Experiment." Plos One 7:8.

Thiessen S, G Gleixner, T Wutzler, and M Reichstein. 2013. "Both Priming and Temperature Sensitivity of Soil Organic Matter Decomposition Depend on Microbial Biomass - an Incubation Study." Soil Biology & Biochemistry 57:739-48.

Thom JGM, GM Dipple, IM Power, and AL Harrison. 2013. "Chrysotile Dissolution Rates: Implications for Carbon Sequestration." Applied Geochemistry 35:244-54.

Todd AS, AH Manning, PL Verplanck, C Crouch, DM McKnight, and R Dunham. 2012. "ClimateChange-Driven Deterioration of Water Quality in a Mineralized Watershed." Environmental Science & Technology 46:9324-32.

Tomkinson T, MR Lee, DF Mark, and CL Smith. 2013. "Sequestration of Martian CO_2 by Mineral Carbonation." Nature Communications 4:6.

Tripathi N, RS Singh, and CP Nathanail. 2014. "Mine Spoil Acts as a Sink of Carbon Dioxide in Indian Dry Tropical Environment." Science of The Total Environment 468–469:1162-71.

Tucker CL, J Bell, E Pendall, and K Ogle. 2013. "Does Declining Carbon-Use Efficiency Explain Thermal Acclimation of Soil Respiration with Warming?" Global Change Biology 19:252-63.

van Groenigen KJ, X Qi, CW Osenberg, Y Luo, and BA Hungate. 2014. "Faster Decomposition under Increased Atmospheric CO_2 Limits Soil Carbon Storage." Science 344:508-09.

Vogel C, CW Mueller, C Hoschen, F Buegger, K Heister, S Schulz, M Schloter, and I Kogel-Knabner. 2014. "Submicron Structures Provide Preferential Spots for Carbon and Nitrogen Sequestration in Soils." Nature Communications 5:7.

von Lutzow M, and I Kogel-Knabner. 2009. "Temperature Sensitivity of Soil Organic Matter Decomposition-What Do We Know?" Biology and Fertility of Soils 46:1-15.

von Lutzow M, I Kogel-Knabner, K Ekschmitt, E Matzner, G Guggenberger, B Marschner, and H Flessa. 2006. "Stabilization of Organic Matter in Temperate Soils: Mechanisms and Their Relevance under Different Soil Conditions - a Review." European Journal of Soil Science 57:426-45.

Vong CQ, N Jacquemet, G Picot-Colbeaux, J Lions, J Rohmer, and O Bouc. 2011. "Reactive Transport Modeling for Impact Assessment of a CO_2 Intrusion on Trace Elements Mobility within Fresh Groundwater and Its Natural Attenuation for Potential Remediation." Energy Procedia 4:3171-78.

Wagai R, AW Kishimoto-Mo, S Yonemura, Y Shirato, S Hiradate, and Y Yagasaki. 2013. "Linking Temperature Sensitivity of Soil Organic Matter Decomposition to Its Molecular Structure, Accessibility, and Microbial Physiology." Global Change Biology 19:1114-25.

Wagai R and LM Mayer. 2007. "Sorptive Stabilization of Organic Matter in Soils by Hydrous Iron Oxides." Geochimica Et Cosmochimica Acta 71:25-35.

Wang S and PR Jaffe. 2004. "Dissolution of a Mineral Phase in Potable Aquifers Due to CO_2 Releases from Deep Formations; Effect of Dissolution Kinetics." Energy Convers. Manage. 45:2833-48.

Wang X, S Piao, P Ciais, P Friedlingstein, RB Myneni, P Cox, M Heimann, J Miller, S Peng, T Wang, H Yang, and A Chen. 2014. "A Two-Fold Increase of Carbon Cycle Sensitivity to Tropical Temperature Variations." Nature 506:212-15.

Washbourne CL, P Renforth, and DAC Manning. 2012. "Investigating Carbonate Formation in Urban Soils as a Method for Capture and Storage of Atmospheric Carbon." Science of The Total Environment 431:166-75.

Wei Y, M Maroto-Valer, and MD Steven. 2011. "Environmental Consequences of Potential Leaks of CO_2 in Soil." Energy Procedia 4:3224-30.

Wieder WR, GB Bonan, and SD Allison. 2013. "Global Soil Carbon Projections Are Improved by Modelling Microbial Processes." Nature Clim. Change 3:909-12.

Wilkin RT and DC DiGiulio. 2010. "Geochemical Impacts to Groundwater from Geologic Carbon Sequestration: Controls on Ph and Inorganic Carbon Concentrations from Reaction Path and Kinetic Modeling." Environ. Sci. Technol. 44:4821-27.

Wilson SA, SLL Barker, GM Dipple, and V Atudorei. 2010. "Isotopic Disequilibrium During Uptake of Atmospheric CO_2 into Mine Process Waters: Implications for CO_2 Sequestration." Environmental Science & Technology 44:9522-29.

Wilson SA, GM Dipple, IM Power, SLL Barker, SJ Fallon, and G Southam. 2011. "Subarctic Weathering of Mineral Wastes Provides a Sink for Atmospheric CO_2." Environmental Science & Technology 45:7727-36.

Wilson SA, GM Dipple, IM Power, JM Thom, RG Anderson, M Raudsepp, JE Gabites, and G Southam. 2009. "Carbon Dioxide Fixation within Mine Wastes of Ultramafic-Hosted Ore Deposits: Examples from the Clinton Creek and Cassiar Chrysotile Deposits, Canada." Economic Geology 104:95-112.

Wilson SA, M Raudsepp, and GM Dipple. 2006. "Verifying and Quantifying Carbon Fixation in Minerals from Serpentine-Rich Mine Tailings Using the Rietveld Method with X-Ray Powder Diffraction Data." American Mineralogist 91:1331-41.

Xia J, J Chen, S Piao, P Ciais, Y Luo, and S Wan. 2014. "Terrestrial Carbon Cycle Affected by Non-Uniform Climate Warming." Nature Geosci 7:173-80.

Zech W, N Senesi, G Guggenberger, K Kaiser, J Lehmann, TM Miano, A Miltner, and G Schroth. 1997. "Factors Controlling Humification and Mineralization of Soil Organic Matter in the Tropics." Geoderma 79:117-61.

Zeigler MP, AS Todd, and CA Caldwell. 2012. "Evidence of Recent Climate Change within the Historic Range of Rio Grande Cutthroat Trout: Implications for Management and Future Persistence." Transactions of the American Fisheries Society 141:1045-59.

Zevenhoven R, J Fagerlund, and JK Songok. 2011. "CO_2 Mineral Sequestration: Developments toward Large-Scale Application." Greenhouse Gases-Science and Technology 1:48-57.

Zhang H-L, R Lal, X Zhao, J-F Xue, and F Chen. 2014. "Chapter One - Opportunities and Challenges of Soil Carbon Sequestration by Conservation Agriculture in China." in Advances in Agronomy, ed. LS Donald, Vol Volume 124, pp. 1-36. Academic Press.

Zheng LG, JA Apps, YQ Zhang, TF Xu, and JT Birkholzer. 2009. "On Mobilization of Lead and Arsenic in Groundwater in Response to CO_2 Leakage from Deep Geological Storage." Chem. Geol. 268:281-97.

Zhou JZ, K Xue, JP Xie, Y Deng, LY Wu, XH Cheng, SF Fei, SP Deng, ZL He, JD Van Nostrand, and YQ Luo. 2012. "Microbial Mediation of Carbon-Cycle Feedbacks to Climate Warming." Nature Climate Change 2:106-10.

In: Climate Change Effects on Soils
Editor: Claudia Holmes

ISBN: 978-1-63482-773-7
© 2015 Nova Science Publishers, Inc.

Chapter 2

CLIMATE CHANGE SCIENCE: KEY POINTS[*]

Jane A. Leggett

SUMMARY

Though climate change science often is portrayed as controversial, broad scientific agreement exists on many points:

- The Earth's climate is warming and changing.
- Human-related emissions of greenhouse gases (GHG) and other pollutants have contributed to warming observed since the 1970s and, if continued, would tend to drive further warming, sea level rise, and other climate shifts.
- Volcanoes, the Earth's relationship to the Sun, solar cycles, and land cover change may be more influential on climate shifts than rising GHG concentrations on other time and geographic scales. Human-induced changes are super-imposed on and interact with natural climate variability.
- The largest uncertainties in climate projections surround feedbacks in the Earth system that augment or dampen the initial influence, or affect the pattern of changes. Feedback mechanisms are apparent in clouds, vegetation, oceans, and potential emissions from soils.

---

[*] This is an edited, reformatted and augmented version of a Congressional Research Service publication, CRS Report for Congress R43229, prepared for Members and Committees of Congress, from www.crs.gov, dated September 10, 2013.

- There is a wide range of projections of future, human-induced climate change, all pointing toward warming and associated sea level rise, with wider uncertainties regarding the nature of precipitation, storms, and other important aspects of climate.
- Human societies and ecosystems are sensitive to climate. Some past climate changes benefited civilizations; others contributed to the demise of some societies. Small future changes may bring benefits for some and adverse effects to others. Large climate changes would be increasingly adverse for a widening scope of populations and ecosystems.

As is common and constructive in science, scientists debate finer points. For example, a large majority but not all scientists find compelling evidence that rising GHG have contributed the most influence on global warming since the 1970s, with solar radiation a smaller influence on that time scale. Most climate modelers project important impacts of unabated GHG emissions, with low likelihoods of catastrophic impacts over this century. Human influences on climate change would continue for centuries after atmospheric concentrations of GHG are stabilized, as the accumulated gases continue to exert effects and as the Earth's systems seek to equilibrate.

The U.S. government and others have invested billions of dollars in research to improve understanding of the Earth's climate system, resulting in major improvements in understanding while major uncertainties remain. However, it is fundamental to the scientific method that science does not provide absolute proofs; all scientific theories are to some degree provisional and may be rejected or modified based on new evidence. Private and public decisions to act or not to act, to reduce the human contribution to climate change or to prepare for future changes, will be made in the context of accumulating evidence (or lack of evidence), accumulating GHG concentrations, ongoing debate about risks, and other considerations (e.g., economics and distributional effects).

BROAD SCIENTIFIC AGREEMENT ON MANY ASPECTS OF CLIMATE CHANGE[1]

Despite portrayals in popular media about controversies in climate change science, almost all climate scientists agree on certain important points:

- The Earth's climate has warmed significantly and changed in other ways over the past century (*Figure 1*). The warming has been widespread but not uniform globally, with most warming over continents at high latitudes, and slight cooling in a few regions, including the southeastern United States, the Amazon, and the North Atlantic south of Greenland.[2]

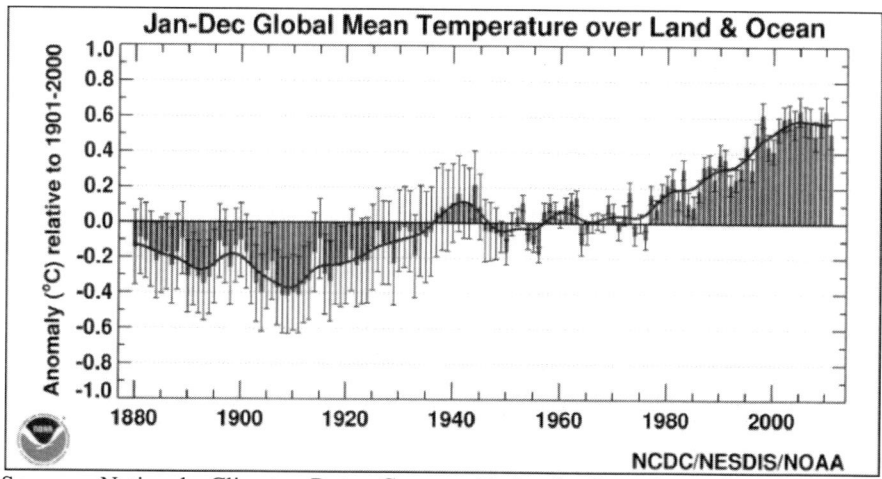

Source: National Climate Data Center, National Oceanic and Atmospheric Administration, U.S. Department of Commerce. Figure extracted March 28, 2013. Very similar findings are reported by several other, independent research groups. See, for example, Rohde, Robert, Richard Muller, Robert Jacobsen, Elizabeth Muller, Saul Perlmutter, Arthur Rosenfeld, Jonathan Wurtele, Donald Groom, and Charlotte Wickham. "A New Estimate of the Average Earth Surface Land Temperature Spanning 1753 to 2011." Geoinformatics & Geostatistics: An Overview 1, no. 1 (December 7, 2012).

Notes: Red bars represent "anomalies," or differences in mean temperature for the year compared with the 20th century average. Anomalies are a better estimate than the absolute value, as they can capture the change over time more reliably while absolute values are vulnerable to gaps in geographic coverage. The blue line shows the running average, applying a "21-point binomial filter" to the time series plotted as red bars. The "whisker" (thin black vertical) lines represent confidence or possible error levels. Confidence has improved over the past century.

Figure 1. Long-Term Temperature Observations. Compared to the 20th Century Global Mean Temperature.

- The climate has varied naturally through geologic history. Past climate changes sometimes proceeded abruptly when they passed certain "tipping points." The National Academy of Sciences concluded that the past few decades were very likely the warmest in the past 400 years, and "that temperatures at many, but not all, individual locations were higher during the past 25 years than during any period of comparable length since A.D. 900."[3] Although conclusions cannot yet be precise, research suggests that global average temperatures today are among the highest since human civilizations began to flourish roughly 4,000 to 8,000 years ago.[4]

- "Greenhouse gases" (GHG) include, among others, carbon dioxide (CO2), water vapor, methane (CH4), and nitrous oxide (N2O), as well as some aerosols. They absorb energy into the atmosphere rather than letting it escape to space. The presence of GHG in the atmosphere warms the Earth to its current temperature.
- Human activities, especially fossil fuel burning, deforestation, agriculture, and some types of industry, have increased GHG concentrations in the atmosphere. CO2, the main GHG emitted by human activities, has risen almost 40% over the past 150 years. About one-third of human-related CO2 has been absorbed by oceans, increasing surface water acidity by 30%.[5]
- The enhanced levels of GHG in the atmosphere contributed to the observed global average warming in recent decades. Over other time and geographic scales, such factors as the Earth's orbit, solar variability, volcanoes, and land cover change have been stronger influences than human-related GHG.
- There is a wide range of projections of future human-induced climate change, with all pointing toward warming. Human-induced change will be superimposed on, and interact with, natural climate variability.
- Human societies and ecosystems are sensitive to climate. Some climate changes benefited civilizations; others contributed to some societies' demises.
- The range of possible impacts on humans and ecosystems is also very wide. In the near term, climate change (including the fertilization of vegetation by CO2) may bring benefits for some, while adversely affecting others. Researchers expect the balance of projected climate change impacts to be increasingly adverse for a widening scope of populations and ecosystems.

As is common and constructive in science, scientists debate finer points. Some disagree with the broader consensus that GHG have been *the major* influence on global warming over the past few decades. Some suggest that, if GHG emissions continue unabated, the resulting climate change would be small and possibly beneficial overall. Most climate modelers project changes that are significant to large, with small likelihoods of changes that could be catastrophic for some human societies and ecosystems in coming decades.

DEALING WITH UNCERTAINTIES

Even the best science cannot provide absolute proof; it is fundamental to the scientific method that all theories are to some degree provisional and may be rejected or modified based on new evidence. Private and public decisions to act or not to act, to reduce the human contribution to climate change or to prepare for future changes, will be made in the context of accumulating evidence (or lack of evidence), accumulating GHG concentrations, ongoing debate about risks, and other considerations (e.g., economics and distributional effects). That said, billions of dollars have been invested in research on a wide range of climate change topics, including the possibility of attribution to alternative causes than greenhouse gases. To date, scientists have found little support for the hypothesis that GHG are not responsible for observed warming, nor have they found much evidence that other factors (including solar changes) can explain more than a small portion of global average temperature increases since the 1970s. For example, measurements of solar irradiance suggest that the solar influence on global temperatures has been decreasing overall since the 1930s, with the up-and-down pattern of the 11-year solar cycle evident in observations. A large body of research is consistent with attributing the majority of global temperature increase since the 1970s to the increase in GHG concentrations. It is this balance of evidence that leads most scientists to consider human-related GHG emissions an important global risk.

Sound Science Does Not Offer Proof

As scientists may point out, "there is no such thing as a scientific proof. Proofs exist only in mathematics and logic, not in science.... The primary criterion and standard of evaluation of scientific theory is evidence, not proof.... The currently accepted theory of a phenomenon is simply the best explanation for it among all available alternatives."6 Normal scientific methods aim at disproving a hypothesis; if evidence cannot disprove a hypothesis, it generally buttresses confidence in the hypothesis.

ISSUES FOR CONGRESS

It appears unlikely that science will provide decision-makers with significantly more scientific certainty for many years regarding the precise patterns and risks of climate change. Nonetheless, both private and public decision-makers face climate-related choices.

Broadly, response options to significant climate change include (1) defer the choices; (2) find out more; (3) inform affected populations; (4) prepare; (5) try to contain it; and (6) choose to experience the consequences. In many cases, many decision-makers are likely to face situations that require a response, such as resolving discrepancies between designated and actual flood plains or attempting to improve agricultural productivity in light of contemporary climate patterns.

Based on what is and what is not well known concerning climate change, as well as other considerations, Members of Congress may address climate-related decisions that affect

- authorizations and appropriations for federal programs, including research and technology development;
- tax and financial incentives for private decision-makers;
- regulatory authorities; or
- information or assistance to affected entities to help them adapt or rebuild after damages.

A variety of other CRS reports provide background and analysis on such options and are listed at the end of this report.[7]

Causes of Observed Climate Change: Forcings, Feedbacks, and Internal Variability

Three concepts may be useful for understanding the mechanisms and debate over the contributions to observed climate change: *forcings, feedbacks*, and *internal variability*.

Forcings

There is broad agreement among scientists that certain factors—including the composition of the atmosphere and solar variability—directly change the balance between incoming and outgoing radiation in the Earth's system and consequently *force* climate change. *Forcings* include the following:

- Atmospheric concentrations of greenhouse gas (GHG) and other trace gas and aerosol. These include water vapor,8 carbon dioxide (CO_2), methane (CH_4), nitrous oxide (N_2O), sulfur hexaflouride (SF6), chlorofluorocarbons (CFC), hydrofluorocarbons (HFC), perfluorocarbons (PFC), ozone, sulfur aerosols, black and organic carbon aerosols, and others. Human activities, especially fossil fuel burning, deforestation, agriculture, and some types of industry, have increased GHG concentrations in the atmosphere. CO_2 has risen almost 40% over the past 150 years.[9]
- Amount and patterns of solar radiation reaching the Earth due to the Earth's orbit around the Sun, and the tilt and wobble of the Earth's axis, as well as solar variability (Figure 2).
- Reflectivity of the Earth's surface due to changes in land use (e.g., urban surfaces, forest cover), changes in ice and snow cover; and vegetation cover.

Human-Related Greenhouse Gas Emissions

A majority of human-related GHG emissions are carbon dioxide, released primarily from energy production and use, deforestation and forest degradation, and cement manufacture. World-wide in 2010, carbon dioxide emissions were 74% of human-related GHG emissions. In the United States, carbon dioxide was 83% of human-related GHG emissions (*Figure 3*). Methane and nitrous oxide emissions are greater shares (16% and 8%, respectively) globally than in the United States (10% and 5%, respectively). Agriculture is a main source of these emissions and is a bigger share of the economies of many low-income countries compared with the United States. Also, many sources in the United States have acted to reduce their GHG emissions (such as in reducing leaks of methane), compared with sources in some low-income countries.

A large majority (73%) of global GHG are emitted by the 10 top emitting countries *Figure 4*). China became the leading GHG emitter in 2007 when it surpassed the United States. While China's emissions have been on the rise, the United States has emitted more cumulatively than any other country over the past 100 years.

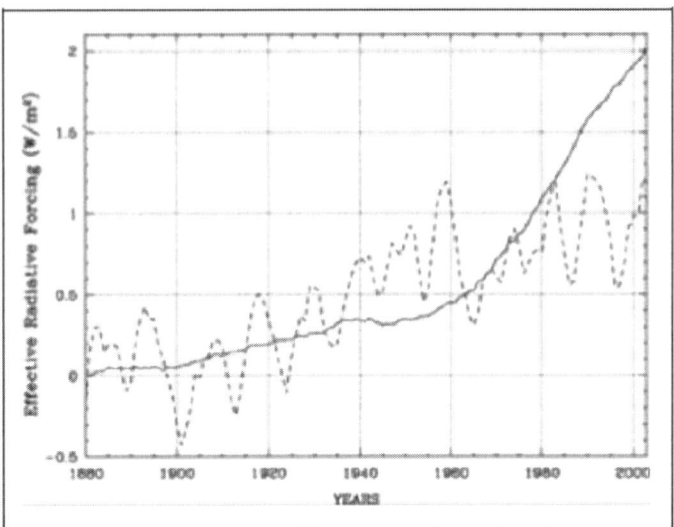

Source: Ziskin, Shlomi, and Nir J. Shaviv. "Quantifying the Role of Solar Radiative Forcing over the 20th Century." Advances in Space Research 50, no. 6 (September 15, 2012): 762–776. doi:10.1016/j.asr.2011.10.009.

Notes: According to the authors, "the optimal anthropogenic contribution (solid line) and the optimal solar contribution (dashed line) over the 20th century. The anthropogenic contribution is primarily composed of GHGs and aerosols. The solar contribution includes changes in the total solar irradiance and the indirect solar effect (ISE)." This is one of many studies, using a variety of methods, investigating the relative contributions of different climate forcings that conclude that the GHG concentrations have outweighed all other influences on global mean air surface temperature from the late 1970s to the present. For a broader, more thorough review of scientific understanding of the solar influence, see Gray, L. J., J. Beer, M. Geller, J. D. Haigh, M. Lockwood, K. Matthes, U. Cubasch, et al. "Solar Influences on Climate." Reviews of Geophysics 48, no. 4 (October 30, 2010). doi:10.1029/2009RG000282.

Figure 2. One Estimate of Human-Related versus Solar Contributions to Global Temperature Change Over the 20t[h] Century.

Feedbacks

Once a change in the Earth's climate system is underway, responses *within* the system will amplify or dampen the initiated change. Virtually all climate scientists conclude that all the feedbacks *in net* are likely to be positive (i.e., increasing climate change in the same direction caused by warming),[10] especially if temperature increases are large;[11] there remain wide differences in views.

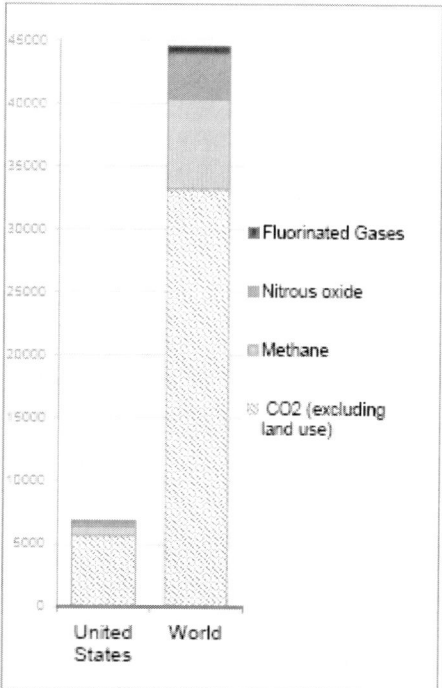

Source: CRS figure using estimates from World Resources Institute, CAIT version 2.0, extracted September 12, 2013.
Notes: These estimates cover the six GHG covered by the Kyoto Protocol (CO_2, CH_4, N_2O, SF_6, HFC, and PFC), expressed in their equivalencies to the effect of CO_2 on "radiative forcing" of the atmosphere over a 100-year period.
The World column includes U.S. emissions.

Figure 3. Shares of Human-Related GHG Emissions by Gas in 2010. Million metric tons of CO_2-equivalent.

An important consideration is that, once positive feedbacks begin, they may be essentially irreversible and, at least theoretically, lead to "runaway warming." A few of the major feedbacks are clouds, vegetation, snow and ice cover, and uptake or releases of GHG by soils and water bodies. Forests, for example, provide both negative and positive feedbacks. On the one hand, higher CO2 concentrations in the atmosphere tend to fertilize their growth (if other conditions are not limiting) and forests may grow more rapidly with greater warmth and precipitation; these factors could dampen initiated warming. On the other hand, forests thrive within certain bounds of growing conditions; if their climate conditions change beyond those bounds, they are likely to grow more slowly and eventually die back, releasing the carbon they and forest soils store and enhancing the initiated climate change.

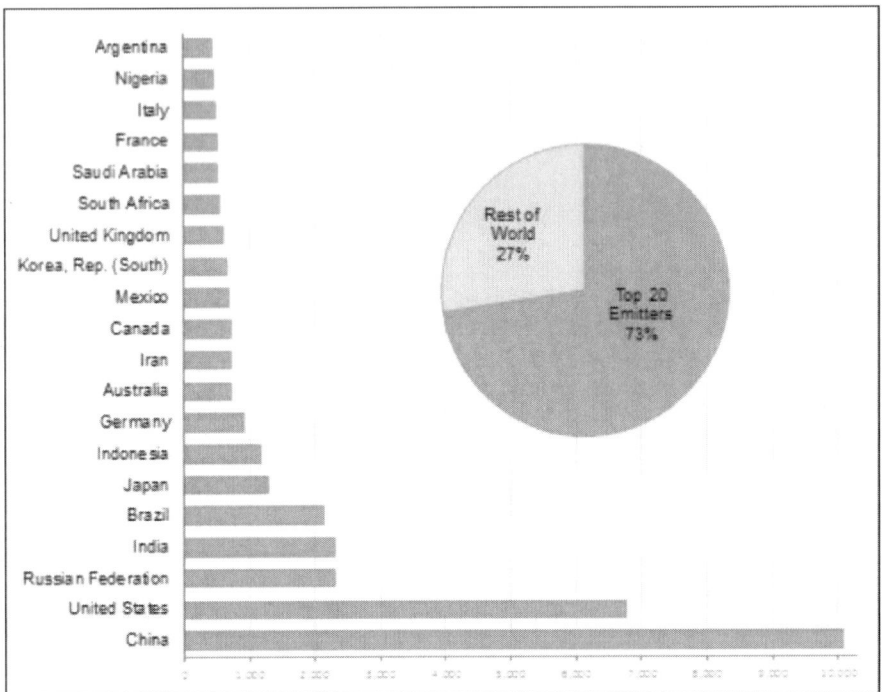

Source: CRS graphic with emission estimates from World Resources Institute, CAIT Version 2.0, extracted September 12, 2013.
Notes: These estimates cover the six GHG covered by the Kyoto Protocol (CO_2, CH_4, N_2O, SF_6, HFC, and PFC), expressed in their equivalencies to the effect of CO_2 on "radiative forcing" of the atmosphere over a 100- year period.

Figure 4. Estimated Top 20 Emitting Nations of Greenhouse Gases in 2010. Million metric tons of carbon-dioxide equivalent, includes all net land use fluxes.

Internal Variability

The climate exhibits its own rhythms, or *internal variability*. The oscillation between El Niño and La Niña events is an example of internal climate variability that has important effects on economies and ecosystems in the Pacific basin (including across the United States). Another is the North Atlantic Oscillation. Internal variability may be difficult to distinguish from decadalscale climate change. Such patterns of variability also may be influenced by climate change.

PROJECTIONS OF FUTURE HUMAN-INDUCED CLIMATE CHANGE

Most climate science experts project that if GHG emissions are not reduced far below current levels, the Earth's climate would warm further, above natural variations, to levels never experienced by human civilizations. If, and as, the climate moves further from its present state, it would reconfigure the patterns and events to which current human and ecological systems are adapted, and the risk of abrupt changes would dramatically increase.

Scenarios of future GHG concentrations under current policies range from 500 ppm carbon dioxide equivalents[12] (CO_2e) to over 1,000 ppm CO_2e by 2100. These are projected to raise the global average temperature by at least 1.5° Celsius (2.7° Fahrenheit) above 1990 levels,[13] not taking into account natural variability. The estimates considered most likely by many scientists are for GHG-induced temperature increases around 2.5 to 3.2° C (4.5 to 5.8° F) by 2100.[14] There is a small but not trivial likelihood that the GHG-induced temperature rise may exceed 6.4°C (11.5° F) above natural variability by 2100.[15]

As context, the global average temperature at the Last Glacial Maximum has been estimated to be about 3 to 5° C (5.4 to 9° F) cooler than present,[16] and is estimated currently to be approaching the highest level experienced since the emergence of human civilizations about 8,000 years ago.[17]

Future climate change may advance relatively smoothly or sporadically, and some regions are likely to experience more fluctuations in temperature, precipitation, and frequency or intensity of extreme events than others. Almost all regions are expected to experience warming; some are projected to become warmer and wetter, while others would become warmer and drier. Sea levels could rise due to ocean warming alone on average between 7 and 23 inches by 2100. Adding to that estimate would be the effects of poorly understood but possible accelerated melting of the Greenland or Antarctic ice sheets. Recent scientific studies have projected a total global average sea level to rise in the 21st century, depending on GHG scenarios, ice dynamics, and other factors, in the range of 2 to 2.5 feet, with a few estimates ranging up to 6.5 feet.[18] Continued warming could lead to additional sea level rise over subsequent centuries of several to many meters. Improving understanding of ice dynamics is a high priority for scientific research to improve sea level rise projections.

Patterns consistent among different climate change models have led to some common expectations: GHG-induced climate change would include more heat waves and fewer extreme cold episodes; more precipitation on average but more droughts in some regions; and generally increased summer warming and dryness in the central portions of continents. Regional changes may vary from the global average changes, however. Scientists also expect precipitation to become more intense when it occurs, thereby increasing runoff and flooding risks.

Precipitation is a particularly challenging component of projecting future climate. For example, for the contiguous United States, recent climate modeling consistently anticipates overall temperature increases, but different models produce a wide range of precipitation changes, from net decreases to net increases.[19] This is particularly problematic in that precipitation, and its characteristics, is closely associated with impacts on agriculture, water supply, streamflows, and other critical systems.

Scientific expectations and model projections consistently point to a global average increase in precipitation with strong variations across regions and time. Generally, dry areas are expected to get dryer, and wet regions are expected to get wetter. In many regions, the increase in evapotranspiration is expected to exceed the increase in precipitation, resulting in general drying of soils and increasing risks of droughts. Precipitation, when it occurs, is expected to be more intense. There will be more energy available for storms, including hurricanes and thunderstorms, though whether they may increase in frequency remains unclear. Sea ice cover in the Arctic is projected to continue its recent decline (*Figure 5*). Greenland is expected to continue ice loss, adding to sea level rise, with more uncertainty about what may happen to ice cover in Antarctica (*Figure 5*). Because Arctic sea ice already floats on water, its melting would not increase sea levels, but large scale melting of land-based ice in Greenland and Antarctica could increase average sea levels by as many as 2 meters by 2100 and several more meters over coming centuries.

IMPACTS OF CLIMATE CHANGE

Nearly every human and natural system could be affected by climate changes, directly or indirectly. The U.S. Global Change Research Program has produced several assessments of scientific understanding of impacts of climate change on the United States.[20]

Climate Change Science: Key Points 73

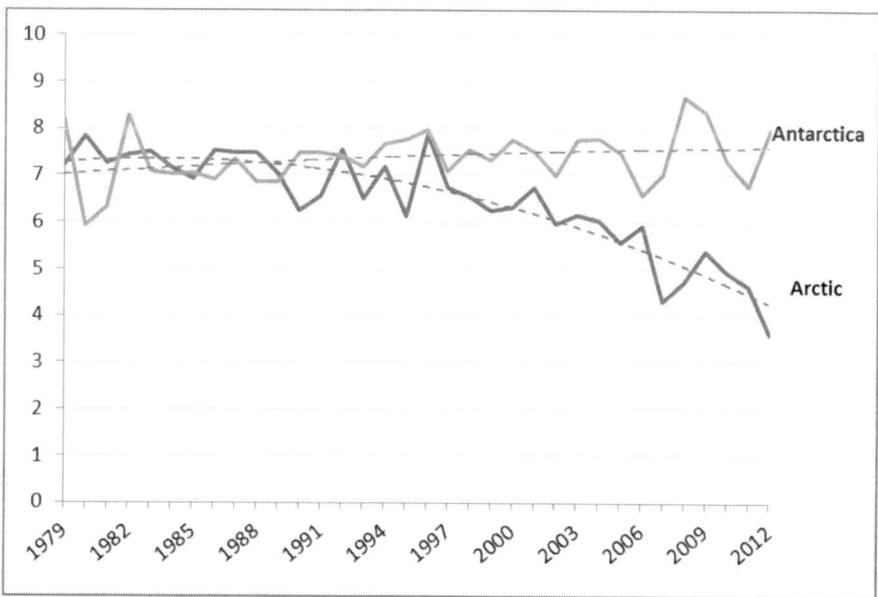

Source: CRS figure from data at the National Snow and Ice Data Center (extracted March 29, 2013), at http://nsidc.org/data/docs/noaa/g02135_seaice_index/ #monthly_graphs_format.

Notes: The dotted lines show the best polynomial fit (2-order) to each time series, as estimated by Excel. For both series, the polynomical fit was slightly better than a linear fit. See the National Snow and Ice Data Center website, referenced above, for further description of the underlying data.

Figure 5. Sea Ice Extent in Arctic (September) and Antarctica (April), 1979 to 2012. (in millions of square kilometers)

Climate Changes Would Affect A Wide Set of Human Systems

Changes in patterns of temperature, precipitation, sea levels, storms, and heat waves (among other indicators of climate) would affect, among other systems:

- water resources and delivery;
- agricultural productivity;
- the frequency and intensity of extreme weather events;
- spread of infectious diseases; air and water pollution levels;
- reliability of transportation, energy, and coastal protection systems;

- commodity prices;
- insurance pay-outs; and
- migration of people and species.

There are many additional elements of the economy and society that could be affected by shifts in climate. Research on potential impacts of climate change is generally less funded and developed than on the climate system itself.

Whether climate changes are meaningful in a policy context arguably depends, on the one hand, on how they influence existing and emerging human systems, and on the other hand, the values people attach to different resources and risks. Past climate changes, often regional not global, contributed to major societal changes, including some large-scale migrations and even the demise of some civilizations.[21] Some climate changes likely stimulated technological advances, such as development of irrigation systems.

Many investments in current buildings, transportation, water systems, agricultural hybrid varieties, and other infrastructure were designed in the context of a climate of one or more decades ago, cooler on average than today. To the degree that climate patterns were factored into design, the investments typically presumed that climate would remain stable within historical bounds of variability. For example, levees may have been built to withstand a 100-year flood (1% chance to occur each year) according to historic runoff, streamflow, and storm surge conditions going back many decades. As climate changes produce greater, more intense precipitation and run-off, however, a 100-year flood may now be closer to the 50-year flood (2% chance flood), and potentially the 10-year flood within decades (10% change flood). If climate continues to change from the conditions for which infrastructure and practices were designed, the risks of losses due to maladaptation would increase.

A wide band of uncertainty surrounds projections of impacts of climate change and, in particular, the critical thresholds for non-linear or abrupt effects. Some impacts of climate change are expected to be beneficial in some locations with a few degrees of warming (e.g., increased agricultural productivity in some regions, less need for space heating, less cold weather mortality, opening of the Northwest Passage for shipping and resource exploitation). Most impacts are expected to be adverse (e.g., lower agricultural productivity in many regions, drought, rising sea levels, spread of disease vectors, greater needs for cooling). Many impacts may be substantial but hard to assess as yet as "positive" or "negative," such as possible impacts on the structure of global financial markets. Risks of abrupt, surprising climate changes with accompanying dislocations are expected to increase as global average temperature increases; some could push natural and socioeconomic systems past key thresholds of tolerance.

Risks of future climate change would be reduced by efforts that reduce vulnerability and build resilience ("adaptation"). Some populations will have the resources to migrate and adapt successfully—even profit from new opportunities that will emerge—while others could lose livelihoods or lives. Adaptations can help mitigate impacts and damage costs, but also impose costs, often on those who can least afford them. Climate change will occur with different magnitudes and characteristics in different regions. The difficulties involved in improving predictions at regional and local scales will challenge preparations for climate change. To a large degree, climate change will expand the uncertainties that individuals and organizations face.

Climate change could have a wide array of effects on individuals, communities, and populations on a large scale. Many of these are expected to occur in small increments: shortages and increasing prices for clean water, rising food prices, higher rates of allergies and such illnesses as diarrhea or cholera, erosion of beaches, etc.

At an increasing rate may be shocks, or distinct weather events, such as more extreme heat waves, severe droughts, or loss of industrial cooling systems when intake water is in short supply or is warmer than tolerable temperatures.[22]

Atmospheric Carbon Dioxide is Increasing Ocean Acidity

The *acidity of the surface waters of the oceans* has increased by about 26% over the past 150 years.[23] Ocean acidification has occurred along with the rise in atmospheric concentrations of CO_2. The oceans remove 25%-40% of the carbon dioxide emissions added annually to the atmosphere by burning fossil fuels. The carbon dioxide absorbed in the oceans decreases the water's pH, an indicator of increasing acidity. According to a National Research Council (NRC) report, the current rate of acidification "exceeds any known change in ocean chemistry for at least 800,000 years."[24] Research shows varying sensitivities of different marine species to rising acidity, making general statements about impacts of ocean acidification difficult.

The NRC concluded,

> While the ultimate consequences are still unknown, there is a risk of ecosystem changes that threaten coral reefs, fisheries, protected species, and other natural resources of value to society. (*Executive Summary*, pp. 3-4)
>
> Congress enacted the Federal Ocean Acidification Research and Monitoring Act of 2009 (P.L. 111-11, Section 12311, Subtitle D) to improve monitoring and research, to assess carbon storage in the oceans and potential effects on acidification and other ocean conditions, and to develop predictive models for future changes in ocean chemistry and marine ecosystems. The program is housed within the National Oceanic and Atmospheric Administration (NOAA), and coordinated with other agencies through an interagency plan through the National Ocean Council.

Extreme events, chronic economic losses, or improved opportunities elsewhere are expected to prompt migration of millions of people, largely within countries, but also across national borders. Extreme events, greater variability, and uncertainty are expected to increase stress and mental health challenges. Some experts project that climate changes could amplify instabilities in countries with weak governance and increase security risks.[25] This may have implications for international political stability and security.[26]

For some experts and stakeholders, likely ecological disruptions (and limitations on species' and habitats' abilities to adapt at the projected rate of climate change) are among the most compelling reasons that humans must act to reduce their interference with the climate system. Some believe humans will have the wherewithal to cope, but non-human systems may not. As the degree and distribution of climate changes continue, ranges of species are likely to change. Climate change is highly likely to create substantial changes in ecological systems and services[27] in some locations, and may lead to ecological surprises.[28] The disappearance of some types of regional ecosystems raises risks of extinctions of species, especially those with narrow geographic or climatic distributions, and where existing ecological communities disintegrate.[29] One set of researchers found "a close correspondence between regions with globally disappearing climates and previously identified biodiversity hotspots; for these regions, standard conservation solutions (e.g., assisted migration and networked reserves) may be insufficient to preserve biodiversity."[30]

SELECTED, RELATED CRS REPORTS

CRS Report R43185, *Ocean Acidification*, by Harold F. Upton and Peter Folger.
CRS Report R41153, *Changes in the Arctic: Background and Issues for Congress*, coordinated by Ronald O'Rourke.
CRS Report RL34580, *Drought in the United States: Causes and Issues for Congress*, by Peter Folger, Betsy A. Cody, and Nicole T. Carter.
CRS Report R43199, *Energy-Water Nexus: The Energy Sector's Water Use*, by Nicole T. Carter.
CRS Report R42611, *Oil Sands and the Keystone XL Pipeline: Background and Selected Environmental Issues*, coordinated by Jonathan L. Ramseur.
CRS Report R42756, *Energy Policy: 113th Congress Issues*, by Carl E. Behrens.
CRS Report R42613, *Climate Change and Existing Law: A Survey of Legal Issues Past, Present, and Future*, by Robert Meltz.
CRS Report R43120, *President Obama's Climate Action Plan*, coordinated by Jane A. Leggett. CRS Report R42756, *Energy Policy: 113th Congress Issues*, by Carl E. Behrens. CRS Report RL34266, *Climate Change: Science Highlights*, by Jane A. Leggett.
CRS Report R41973, *Climate Change: Conceptual Approaches and Policy Tools*, by Jane A. Leggett.

End Notes

[1] This CRS report will be reviewed and, as appropriate, revised considering evidence provided from emerging scientific research. Of, note, the Intergovernmental Panel on Climate Change (IPCC) will release its fifth assessment report later in 2013.
[2] For more information, see maps available at the National Climate Data Center, http://www.ncdc.noaa.gov/oa/climate/globalwarming.html and http://www.ncdc.noaa.gov/oa/climate/research/trends.html#global.
[3] Board on Atmospheric Sciences and Climate. Surface Temperature Reconstructions for the Last 2,000 Years. National Research Council, 2006. http://books.nap.edu/openbook.php?record_id=11676&page=1.
[4] Marcott, Shaun A., Jeremy D. Shakun, Peter U. Clark, and Alan C. Mix. "A Reconstruction of Regional and Global Temperature for the Past 11,300 Years." Science 339, no. 6124 (March 8, 2013): 1198–1201. doi:10.1126/science.1228026; Kellerhals, T., S. Brütsch, M. Sigl, S. Knüsel, H. W. Gäggeler, and M. Schwikowski. "Ammonium Concentration in Ice Cores: A New Proxy for Regional Temperature Reconstruction?" Journal of Geophysical Research: Atmospheres 115, no. D16 (2010): n/a–n/a. doi:10.1029/2009JD012603; Thibodeau, Benoît, Anne de Vernal, Claude Hillaire-Marcel, and Alfonso Mucci.

"Twentieth Century Warming in Deep Waters of the Gulf of St. Lawrence: A Unique Feature of the Last Millennium." Geophysical Research Letters 37, no. 17 (2010): n/a–n/a. doi:10.1029/2010GL044771. See also the references at http://www.globalwarmingart.com/wiki/File:Holocene_Temperature_Variations_Rev_png, which depict a collection of major temperature reconstructions of the Holocene, as well as the broad range of uncertainty of available estimates and the average of those estimates.

[5] National Research Council. Ocean Acidification: A National Strategy to Meet the Challenges of a Changing Ocean. Washington DC, 2013; Feely, Richard A. 2010. A Rational Discussion of Climate Change: The Science, the Evidence, the Response. Testimony before the House Committee on Science and Technology, Subcommittee on Energy and Commerce. Washington DC. (p.130). See also CRS Report R43185, Ocean Acidification, by Harold F. Upton and Peter Folger.

[6] See, for example, the discussion in Kanazawa, Satoshi. "Common misconceptions about science I: "Scientific proof"." Psychology Today, November 16, 2008. http://www.psychologytoday.com/blog/the-scientific-fundamentalist/ 200811/common-misconceptions-about-science-i-scientific-proof.

[7] Many CRS reports related to climate change may be found at Issues Before Congress: Climate Change Science, Technology, and Policy, at http://www.crs.gov/pages/subissue.aspx?cliid=3878&parentid=2522&preview=False.

[8] Water vapor is the most important GHG in the atmosphere but is understood not to be directly influenced by humans; it would be, however, involved in feedback mechanisms, discussed later.

[9] About one-third of human-related CO2 has been absorbed by oceans, increasing surface water acidity by 30%. See National Research Council. Ocean Acidification: A National Strategy to Meet the Challenges of a Changing Ocean. Washington, DC, 2013; Feely, Richard A. 2010. A Rational Discussion of Climate Change: The Science, the Evidence, the Response. Testimony before the House Committee on Science and Technology, Subcommittee on Energy and Commerce. Washington, DC. (p.130).

[10] One line of evidence is that carbon dioxide levels have varied closely with the Earth's temperature in and out of glacial periods over the past million years. These cycles are mostly triggered by changes in the Earth's orbit, tilt, and wobble. In some of these cycles, temperatures rose in advance of rising atmospheric carbon dioxide concentrations. Scientists generally interpret this as a tendency for positive climate warming feedbacks that increase carbon dioxide concentrations which then enhance warming, etc.—that the net positive feedbacks amplify an initial climate warming.

[11] Positive feedbacks could increase if and when, for example, large tracts of forests die back as a response to exceeding their climate thresholds of tolerance, or current permafrost thaws and releases the carbon it contains, or if reservoirs of methane hydrates destabilize.

[12] In order to show multiple gases of different potencies on a single scale, GHG have been indexed relative to the effect that a mass of CO2 would have over several time periods (because GHG remain in the atmosphere for different lengths of time, from days to tens of thousands of years). The index used for these estimates uses a 100-year time horizon, the most frequently used period.

[13] Intergovernmental Panel on Climate Change Working Group I, Climate Change 2007: The Physical Basis (Cambridge, UK: Cambridge University Press, 2007).

[14] As a point of reference, the global mean annual temperature during the 20th century is estimated to have been approximately 13.9o Celsius (57.0o Fahrenheit), according to NOAA's National Climate Data Center.

15 Ibid.

16 Intergovernmental Panel on Climate Change Working Group I. Climate Change 2007: The Physical Basis. Cambridge, UK: Cambridge University Press, 2007. Executive Summary.

17 Highest temperatures of the Holocene may have occurred in one or more periods some 5,000 to 8,000 years ago, although sufficient data are not available for all parts of the globe to have reliable estimates of average global temperature. The oldest cities discovered date from approximately the same period, such as the extensive settlement of Byblos in present-day Lebanon, by about 6,000 years ago, or Medinat Al-Fayoum in Egypt, about 6,000 years old. Since the early to mid-Holocene, however, average temperatures appear to have been declining slowly, with notable periods of warming and cooling. The changes entailed in Holocene climate variability have been significant in terms of effects on humans and ecosystems, and have led to both benefits to, and the demise of, numerous civilizations.

18 See discussion in National Research Council. Advancing the Science of Climate Change. Washington DC, 2010, at p. 244.

19 See, for example, climate change scenarios available from the U.S. Global Change Research Program at http://scenarios.globalchange.gov/sites/default/ files/b/figures/ UnitedStates/ Ann_US_precip_a2.png, with notes at http://scenarios.globalchange.gov/node/1087. See also discussion in this report regarding dealing with uncertainties.

20 Karl, Thomas R., Mellillo, Jerry M., and Peterson, Thomas C. (eds.) Global Change Impacts in the United States. U.S. Global Change Research Program. 2009. Such periodic assessments are required by the Global Change Research Act of 1990 (P.L. 101-606). A new national assessment of impacts on the United States is due in late 2013.

21 There is a growing set of research on the relationship between past climate change and civilizations. A sample of recent research includes Buckley, B. M., K. J. Anchukaitis, D. Penny, R. Fletcher, E. R. Cook, M. Sano, L. C. Nam, A. Wichienkeeo, T. T. Minh, and T. M. Hong. "Climate as a Contributing Factor in the Demise of Angkor, Cambodia." Proceedings of the National Academy of Sciences 107, no. 15 (March 2010): 6748–6752; Cook, Edward R, Kevin J Anchukaitis, Brendan M Buckley, Rosanne D D'Arrigo, Gordon C Jacoby, and William E Wright. "Asian Monsoon Failure and Megadrought During the Last Millennium." Science (New York, N.Y.) 328, no. 5977 (April 23, 2010): 486– 489; DeMenocal, PB. "Cultural Responses to Climate Change During the Late Holocene." Science (Washington) 292, no. 5517 (April 27, 2001): 667–673; Haug, G. H, D. Gunther, L. C Peterson, D. M Sigman, K. A Hughen, and B. Aeschlimann. "Climate and the Collapse of Maya Civilization." Science 299, no. 5613 (2003): 1731; Scholz, Christopher A., Thomas C. Johnson, Andrew S. Cohen, John W. King, John A. Peck, Jonathan T. Overpeck, Michael R. Talbot, et al. "East African Megadroughts Between 135 and 75 Thousand Years Ago and Bearing on Early-modern Human Origins." Proceedings of the National Academy of Sciences of the United States of America 104, no. 42 (October 16, 2007): 16416–16421.

22 For examples of these risks to power plants, see Department of Energy (DOE), U.S. Energy Sector Vulnerabilities to Climate Change and Extreme Weather. July 2013. http://energy.gov/sites/prod/files/2013/07/f2/20130710-EnergySector-Vulnerabilities-Report.pdf. See also CRS Report R43199, Energy-Water Nexus: The Energy Sector's Water Use, by Nicole T. Carter.

23 NRC Committee on the Development of an Integrated Science Strategy for Ocean Acidification Monitoring, Research, and Impacts Assessment; National Research Council. "Executive Summary." In Ocean Acidification: A National Strategy to Meet the Challenges of a Changing Ocean. Prepublication. Washington, D.C.: The National Academies Press, 2010; Jacobson, Mark Z. "Studying ocean acidification with conservative, stable numerical

schemes for nonequilibrium air-ocean exchange and ocean equilibrium chemistry." Journal of Geophysical Research 110 (April 2, 2005): 17 PP.

[24] Ibid.

[25] An example of this is the adverse weather events in early 2011 that led to spikes in food prices and contributed to demonstrations in Tunisia and Egypt. These, in turn, led to regime change, although one cannot attribute these events to climate change, as opposed to weather variability, and the political implications might have been very different in regimes with better economic performance, less income disparity, fewer allegations of corruption, and greater social resilience. The point remains, nonetheless, that societies are sensitive to climatic variables in many ways.

[26] Regarding risks to national security, see, for example, Defense Science Board Task Force on Trends and Implications of Climate Change for National and International Security. October 2011. http://www.acq.osd.mil/dsb/ reports/ADA552760.pdf; and U.S. Department of Defense. Quadrennial Defense Review, February 2010. (pp. xv, 84- 88) http://www.defense.gov/qdr/qdr%20as%20of%2029jan10%201600.pdf.

[27] Economists and scientists sometimes refer to "ecosystem services," which are the services that natural systems provide and for which, very frequently, humans do not typically pay. Ecosystems services include water filtration, filtering of air pollution, recreational and spiritual opportunities, etc. Even without being valued in capital markets, ecosystem services may be critically important to economies. For example, in many coastal areas, mangroves or wetlands provide valuable buffering against frequent storm and flood events. If such ecosystem services did not exist, communities would have to pay for manufactured alternatives (e.g., sea walls) or risk incurring damages.

[28] For example, the very rapid spread of pine beetles in recent years was unexpected and caused large damages (although a temporarily inexpensive supply of timber) in a very short period. See CRS Report R40203, Mountain Pine Beetles and Forest Destruction: Effects, Responses, and Relationship to Climate Change.

[29] Malcolm, Jay R., Canran Liu, Ronald P. Neilson, Lara Hansen, and Lee Hannah. "Global Warming and Extinctions of Endemic Species from Biodiversity Hotspots." Conservation Biology 20, no. 2 (2006): 538-548.

[30] John W. Williams, Stephen T Jackson, and John E. Kutzbach, "Projected distributions of novel and disappearing climates by 2100 AD," Proceedings of the National Academy of Sciences of the United States of America 104, no. 14 (April 3, 2007).

INDEX

#

20th century, 63, 68, 78
21st century, 36, 71

A

abiotic C cycling, vii
abstraction, 29
access, 30
accounting, 13
acetic acid, 17
acid, 5, 12, 13, 14, 17, 25, 30, 35
acidic, vii, 14, 30, 33
acidity, 6, 64, 75, 78
activation energy, 10, 37
AD, 80
adaptation, 75
adhesive strength, 34
adsorption, 13, 31, 33, 34, 35
advancement, 14
adverse effects, viii, 62
adverse weather, 80
aerosols, 64, 67, 68
age, 9, 25, 29
agencies, 75
aggregation, 21
agriculture, 1, 3, 15, 64, 67, 72
air temperature, 1
Alaska, 55
alkalinity, 5, 13, 17
alternative causes, 65
aluminum oxide, 35
amino, 26
amino acid(s), 26
appropriations, 66
aquatic life, 14
aromatic compounds, 29, 40
assessment, 77, 79
atmosphere, 2, 3, 4, 8, 9, 12, 13, 16, 18, 20, 21, 22, 36, 64, 66, 67, 69, 70, 75, 78
attribution, 65
authorities, 66

B

beetles, 80
benefits, viii, 62, 64, 79
BI, 48
bicarbonate, vii, 7, 8, 9, 10, 13, 16
biodiversity, 76
biological activity, 21
biological behavior, 18
biomass, 12, 37
biopolymers, 20, 28
biosphere, 2, 8
biotic, 4, 29
bonding, 33, 34
bonds, 30
bounds, 69, 74

Index

breathing, 8

C

Ca^{2+}, vii, 10
Cambodia, 79
candidates, 9
capital markets, 80
carbohydrate(s), 26, 31
carbon, vii, 1, 2, 19, 64, 67, 70, 71, 75, 78
carbon dioxide, 1, 64, 67, 71, 75, 78
carbonate, vii, 2, 4, 7, 8, 9, 10, 11, 12, 13, 16, 17
carboxyl, 31
catchments, 16
cation, 7, 34
cations, vii, 7, 13, 16, 35
cellulose, 20
challenges, 76
charge density, 10
chemical, 2, 3, 5, 7, 8, 13, 14, 15, 18, 19, 20, 25, 28, 29, 30, 33, 34, 35, 36, 39, 40
chemical bonds, 33, 34
chemical characteristics, 36
chemical interaction, 29
chemical properties, 29
chemical reactions, 39, 40
China, 20, 43, 53, 58, 67
chlorophyll, 20
cholera, 75
cities, 79
civilizations, viii, 62, 64, 71, 74, 79
clay minerals, 5, 17, 33
climate change, vii, viii, ix, 2, 3, 4, 6, 11, 12, 14, 18, 20, 21, 22, 23, 25, 27, 28, 34, 35, 36, 40, 61, 62, 63, 64, 65, 66, 67, 68, 70, 71, 72, 73, 74, 75, 76, 78, 79, 80
climate extremes, 3
climates, 76, 80
climatic factors, 27
clouds, viii, 61, 69
clusters, 32
CO2, vii, 1, 4, 5, 6, 7, 8, 9, 10, 11, 12, 13, 15, 16, 17, 18, 20, 21, 22, 36, 37, 41, 42, 44, 45, 46, 47, 48, 49, 50, 51, 52, 54, 56, 57, 58, 64, 67, 69, 70, 75, 78
combined effect, 21, 23
commodity, 73
communities, 19, 75, 76, 80
community, 15, 29, 37
competition, 12
complexity, 33
composition, 6, 9, 13, 20, 21, 28, 33, 40, 66
compounds, 20, 21, 29, 30, 31, 33, 35, 39
conceptual model, 36
configuration, 33
Congo, 16
Congress, 61, 66, 75, 77, 78
connectivity, 30
consensus, 36, 64
conservation, 22, 76
consumption, 7, 8, 10
contaminant, 5, 6
contamination, 12
Continental, 41, 56
controversial, vii, 61
controversies, 62
cooling, 63, 74, 75, 79
coral reefs, 75
correlation(s), 31, 32
corruption, 80
covering, 6
crust, 8
crystal growth, 35
crystallinity, 6
CT, 42, 53
cultivation, 27
cutin, 20
CV, 54
cycles, viii, 1, 2, 4, 8, 15, 26, 61, 78
cycling, vii, 2, 3, 5, 15, 18, 25, 26, 29

D

damages, 66, 80
danger, 14
decay, 37
decomposition, 11, 21, 23, 26, 27, 28, 29, 30, 33, 36, 37, 38, 39, 40

deforestation, 64, 67
degradation, 19, 20, 67
dehydration, 10
demonstrations, 80
Department of Defense, 80
Department of Energy, 1, 79
depolymerization, 38
deposits, 2
depth, 6, 19, 25, 27, 40
desorption, 30, 33
diarrhea, 75
diatoms, 17
differential scanning, 29
differential scanning calorimetry, 29
diffusion, 12, 30
diseases, 73
dispersion, 25
dissociation, 5
distribution, 5, 19, 22, 25, 26, 27, 30, 76
diversity, 20
DOC, 2, 27
drainage, 12, 14
drought, 1, 4, 14, 22, 74
drying, 8, 14, 72
DSC, 29

E

ecological systems, 71, 76
ecology, 29
economic losses, 76
economic performance, 80
economics, ix, 62, 65
ecosystem(s), viii, 1, 2, 3, 10, 13, 14, 15, 22, 23, 24, 36, 62, 64, 65, 70, 75, 76, 79, 80
Egypt, 79, 80
El Niño, 70
emission, 2, 70
encapsulation, 28, 30
energy, 3, 26, 64, 67, 72, 73, 79
engineering, vii, 12, 16, 17
environment(s), 9, 11, 20, 26, 37
environmental conditions, 7, 21
environmental quality, 1, 4
enzyme(s), 19, 29, 30, 38

equilibrium, 23, 80
erosion, 6, 75
Europe, 39
evapotranspiration, 72
evidence, viii, ix, 15, 24, 30, 62, 65, 77, 78
evolution, 4, 19
exclusion, 32
exploitation, 74
exposure, 5
extraction, 17
extreme cold, 72
extreme weather events, 73

F

farmland, 16
fertility, 18
fertilization, 64
fertilizers, 22
fiber, 21
films, 35
filtration, 80
financial, 66, 75
financial incentives, 66
financial markets, 75
fish, 14
fisheries, 75
fixation, 12, 16
flooding, 72
fluctuations, 71
food, 1, 3, 4, 15, 21, 75, 80
food production, 15
food security, 1, 4
force, 34, 67
formation, 4, 5, 7, 8, 9, 10, 11, 12, 15, 16, 17, 18, 19, 21, 29, 30, 34
freshwater, 13
frost, 1

G

GHG, viii, ix, 61, 62, 64, 65, 67, 68, 69, 70, 71, 72, 78
Global Change Research Act, 79

global climate change, 38
global scale, 4, 20
global warming, viii, 1, 4, 6, 18, 20, 22, 28, 36, 39, 62, 64
glucose, 35, 38
governance, 76
grasses, 15, 31
grasslands, 24
greenhouse, viii, 18, 22, 27, 61, 65, 67
greenhouse gas (GHG), viii, 18, 22, 27, 61, 65, 67
greenhouse gas emissions, 18, 22, 27
greenhouse gases, viii, 61, 65
groundwater, 1, 3, 13, 14
growth, 38, 69

H

habitats, 76
hardwoods, 15
harvesting, 15
Hawaii, 25, 31
HE, 42, 48
hemicellulose, 20
heterogeneity, 8, 25, 34
heterogeneous systems, 34
history, 63
HM, 54
Holocene, 78, 79
host, 21
hotspots, 76
House, 78
human, viii, ix, 2, 3, 62, 64, 65, 67, 71, 72, 74, 76, 78
human health, 3
humidity, 8
hurricanes, 72
hybrid, 74
hydrogen, 11, 33, 34
hydrologic regime, 14
hydrolysis, 11
hydrophobicity, 28, 30
hydrophyte, 10
hydroxyl, 35
hypothesis, 24, 29, 32, 37, 65

I

identity, 21, 32
Impact Assessment, 57
improvements, viii, 62
income, 67, 80
incubation time, 40
India, 12
individuals, 75
Indonesia, 44
industry, 64, 67
infrared spectroscopy, 35
infrastructure, 74
inhibition, 10
inorganic C, vii, 2
interface, 34, 35
interference, 76
interrogations, 19
investments, 74
ions, 10, 16, 17, 34
irrigation, 74
isolation, 19
issues, vii, 3, 12, 17, 19, 22, 25, 28, 38

J

Java, 44
Jordan, 54

K

kinetics, 19, 29
Kyoto Protocol, 69, 70

L

lakes, vii, 5, 13, 14, 15, 17
land cover, viii, 21, 61, 64
landscapes, 2, 6
layered double hydroxides, 8
leaching, 4
lead, 9, 12, 69, 71, 76
leakage, 16

Index

leaks, 67
Lebanon, 79
levees, 74
ligand, 34
light, 9, 66
lignin, 20, 31, 33, 35
lipids, 20
liquid phase, 9
Luo, 49, 56, 58

M

magnesium, 11
magnitude, 8, 9, 11, 14, 15, 38
majority, viii, 39, 62, 65, 67
management, 14, 21, 22, 24, 26, 27
mangroves, 80
manipulation, 15
mantle, 7
marine environment, 27
marine species, 75
Mars, 8
mass, 78
materials, 5, 10, 34
mathematics, 65
matrix, 25, 29, 37
matter, 33, 38, 40
MB, 43
measurements, 40, 65
media, 62
melting, 71, 72
mental health, 76
metal ion(s), 28, 31
metals, 14, 33
meter, 20, 24
Mg^{2+}, 10
micrometer, 34
microorganisms, 10, 30, 33
migration, 73, 76
mine soil, 12
mineralization, vii, 2, 3, 11, 18, 19, 26, 28, 33, 36, 37, 38
misconceptions, 78
models, 8, 19, 23, 24, 25, 29, 36, 72, 75
moisture, 40

moisture content, 40
mole, 7
molecular structure, 28
molecular weight, 10
molecules, 18, 27, 28, 29, 34, 35
monomers, 35
Montana, 48
mortality, 74
MR, 44, 53, 56

N

NaCl, 11
nanometer, 20
nanometer scale, 20
National Academy of Sciences, 46, 47, 48, 54, 63, 79, 80
national borders, 76
National Research Council, 75, 77, 78, 79
national security, 80
National Strategy, 78, 79
natural resources, 75
negative effects, 4
neutral, 35
nitrogen, 22
nitrous oxide, 64, 67
NOAA, 75, 78
nonequilibrium, 80
North America, 8, 51
NRC, 75, 79
nuclear magnetic resonance, 29, 35, 40
nucleation, 10
nuclides, 6
nutrients, 3, 4

O

occlusion, 15, 28, 30
oceans, vii, viii, 2, 5, 13, 14, 61, 64, 75, 78
OH, 11, 31
oil, 3
opportunities, 75, 76, 80
orbit, 64, 67, 78
organic carbon, vii, 2, 19, 67

organic compounds, 2, 20, 29, 33, 34, 35, 39
organic matter, vii, 2, 3, 10, 18, 23, 30, 38, 39
organism, 20
oscillation, 70
overlap, 26
oxidation, 26, 29
oxygen, 11
ozone, 67

P

Pacific, 1, 70
pathways, 10, 20, 28
permafrost, 20, 78
pH, 5, 6, 11, 13, 14, 16, 75
photosynthesis, 10, 13
physicochemical characteristics, 23
physiology, 19, 37
plant growth, 22
plants, 4, 15, 17
PM, 46, 48, 50
polar, 34
policy, 74
pollutants, viii, 12, 61
pollution, 73, 80
polymers, 28
polyphenols, 20
polysaccharide, 33
polysaccharides, 20, 35
pools, 2, 5, 19, 21, 22, 23, 24, 29, 39
population, 14, 21
porosity, 6
positive feedback, 21, 69, 78
power plants, 79
precipitation, viii, 9, 10, 11, 12, 16, 17, 22, 27, 35, 62, 69, 71, 72, 73, 74
preservation, 25, 27, 28, 29, 31
President, 77
President Obama, 77
priming, 37, 38
probe, 19, 30
profit, 75
project, viii, 19, 62, 65, 71, 76

protection, vii, 2, 3, 15, 19, 21, 28, 29, 30, 31, 32, 33, 37, 73
protective role, 30
pumps, 13
pyrite, 14

Q

quartz, 17

R

radiation, viii, 62, 67
rainfall, 1, 4, 25, 26
reactants, 9
reactions, 3, 4, 5, 14, 18, 32, 38
reagents, 30
recommendations, 3
recreational, 80
redistribution, 22
relevance, 18, 25, 28
reliability, 73
researchers, 9, 16, 21, 24, 32, 33, 35, 76
reserves, 76
residues, 20
resilience, 75, 80
resistance, 28, 29
resources, 14, 21, 74, 75
respiration, 18, 21, 37, 40
response, 3, 8, 9, 11, 14, 18, 21, 22, 23, 26, 36, 38, 39, 40, 66, 78
RH, 43, 47
Richland, 19
risk(s), ix, 5, 62, 65, 66, 71, 72, 74, 75, 76, 79, 80
roots, 26
roughness, 6
runoff, 4, 72, 74
Russia, 9, 54

S

salinity, 11
saturation, 16, 17, 23, 24

science, vii, viii, 3, 19, 61, 62, 64, 65, 66, 71, 77, 78
scientific method, viii, 62, 65
scientific theory, 65
scientific understanding, 68, 72
scope, viii, 62, 64
sea level, viii, 61, 62, 71, 72, 73, 74
security, 3, 76
sediments, 13, 27, 32, 34
seeding, 10
self-organization, 31
sensitivity, vii, 3, 18, 19, 27, 29, 36, 37, 38, 39
services, 80
short supply, 75
showing, 26
Siberia, 55
silica, 11, 15, 17
silicate, vii, 2, 6, 7, 8, 9, 11, 15, 16, 17, 33
society, 73, 75
soil minerals, vii, 3, 5, 18, 31, 32, 35
soil particles, 31
soil properties, vii, 1, 4, 18
soil type, 3, 6, 27
solar cycles, viii, 61
solar radiation, viii, 62, 67
solution, 9, 12, 31, 32, 34
SOM, vii, 2, 3, 10, 18, 19, 20, 21, 22, 23, 24, 25, 26, 27, 28, 29, 30, 31, 32, 33, 34, 35, 36, 37, 38, 39, 40
sorption, 24, 30, 31, 32, 33
sorption experiments, 24
SP, 58
Spain, 41
species, 13, 27, 73, 75, 76
specific surface, 32
spectroscopic techniques, 19
spectroscopy, 29, 40
SS, 43, 54
stability, 2, 10, 24, 26, 28, 29, 30, 32, 33, 34, 76
stabilization, 21, 25, 26, 28, 29, 31, 32, 33, 40
stakeholders, 76
starch, 20
state, 3, 19, 23, 29, 71
stock, 23, 39
stoichiometry, 37
storage, 9, 12, 16, 18, 22, 23, 24, 25, 75
storms, viii, 1, 62, 72, 73
stress, 76
structure, 6, 11, 19, 29, 35, 37, 75
suberin, 20
substrate(s), 26, 30, 37, 38
succession, 12
sulfate, 12
sulfur, 67
Sun, viii, 61, 67
surface area, 6, 31, 32
surface layer, 3
surface properties, 32, 33
surfactant, 31
survival, 14

T

tannins, 20
Task Force, 80
techniques, 27, 31
technological advances, 18, 74
technology, 66
TEM, 19
temperature, vii, 1, 3, 4, 6, 9, 10, 14, 18, 19, 22, 26, 29, 36, 37, 38, 39, 40, 63, 64, 65, 68, 71, 72, 73, 75, 78, 79
temperature dependence, 39
terrestrial ecosystems, 13, 15
texture, 6, 27, 32
thermodynamics, 38
time periods, 78
time series, 63, 73
transformation(s), vii, 2, 3, 7, 10, 12, 18, 19, 20
transmission, 19
transmission electron microscopy, 19
transport, 13
transportation, 25, 73, 74
triggers, 15
tropical forests, 26
turnover, 22, 24, 26, 27, 29, 39, 40

U

U.S. Department of Commerce, 63
UK, 78, 79
uniform, 2, 39, 63
United Kingdom, 21
United States, 46, 47, 54, 63, 67, 70, 72, 77, 79, 80
urban, 12, 67
USA, 43, 52

V

vacancies, 8
vapor, 78
variables, 3, 6, 7, 10, 12, 15, 18, 19, 30, 34, 36, 80
variations, 9, 71, 72
varieties, 74

vegetation, viii, 15, 20, 27, 61, 64, 67, 69
Volcanoes, viii, 61
vulnerability, 36, 75

W

Washington, 19, 42, 78, 79
waste, 12
water, 1, 2, 4, 5, 8, 11, 12, 13, 14, 18, 22, 35, 64, 67, 69, 72, 73, 74, 75, 78, 80
water quality, 13, 14
water resources, 14, 18, 73
water supplies, 1, 3
water vapor, 64, 67
weathering, vii, 2, 3, 4, 6, 7, 8, 9, 11, 12, 13, 14, 15, 16, 17, 22
Western Australia, 12
wetlands, 20, 80
worldwide, 16, 24